AIRMANSHIP & AIR LAW

AIRMANSHIP & AIR LAW

RICHARD T. BOWYER

Airlife
England

First published in the UK in 1997
by Airlife Publishing Ltd

British Library Cataloguing-in Publication Data
A catalogue record for this book
is available from the British Library

ISBN 1 85310 376 4

Typeset by Phoenix Typesetting, Ilkley, West Yorkshire.
Printed in England by Livesey Ltd, Shrewsbury.

Airlife Publishing Ltd
101 Longden Road, Shrewsbury, SY3 9EB, England.

Foreword

This book is written as a guide to the professional and private pilot, and in no way compensates for the information contained in official publications which are amended at frequent intervals.

As a guide it contains many of the aspects required for the Air Law examinations of the Civil Aviation Authority and many references, and at the time of publication it includes the latest amendments possible and references for the necessary information.

CONTENTS

CHAPTER 1
CONTROL OF AIRSPACE

Introduction to Air Traffic Control

To avoid and reduce aircraft accidents, a system of aircraft traffic control must be used for the safe separation of aircraft, both in the air and on the ground. To achieve this an Air Traffic Control (ATC) system is used which covers the local area around airports and also the skies of the world. A system is used that covers controlled and uncontrolled airspace, and a set of rules is required to make both types of airspace safe for all air travellers, both military and civilian. The world is divided into regions of control which are called Flight Information Regions (FIRs), and these are under the control of Air Traffic Control Centres (ATCC). These are further divided into local areas of control around civilian airports and military aerodromes.

Terminology

AIRPORT/AERODROME/AIRFIELD An area on the ground or on water that is used or intended to be used by different aircraft for:

a Surface movement (taxying)
b Departure (take-off)
c Arrival (landing).

Aerodromes can be classified under three headings, these being Government Aerodromes, Licensed Aerodromes and Unlicensed Aerodromes. These will later be discussed in more detail.

APRON/TARMAC/DISPERSAL A defined area on a land aerodrome for the purpose of allowing parking, the loading and unloading of passengers and/or cargo, and for refuelling.

AIR TRAFFIC CONTROL A service operated by an authority to promote the safe, orderly and efficient flow of air traffic.

HOLDING POINT/BAY A defined area on a taxi area where an aircraft can be stopped while awaiting for the line-up, or bypassed to allow the efficient surface movement of aircraft without affecting aircraft landing or taking off.

RUNWAY A defined area, either paved or unpaved, provided for the take-off and landing run of aircraft.

TAXIWAY A defined path on a land aerodrome selected or prepared for the taxying of an aircraft. Each taxiway is indicated by a letter of the alphabet, i.e. A – Alpha, B – Bravo, C – Charlie, etc, on a marker board.

MANOEUVRING AREA That part of an aerodrome provided for the take-off and landing of aircraft, and for the movement of aircraft on the surface, excluding the apron and any part of the aerodrome provided for the maintenance of aircraft.

THRESHOLD The beginning of that portion of runway usable for landing.

DISPLACED THRESHOLD A permanent or temporary landing point along the runway, other than the beginning of the runway, established because of work in progress or for some other reason.

Air Traffic Control

Air Traffic Control has the authority and responsibility to exercise control of air traffic on the ground at airports and in the controlled airspace. It has the authority to approve or disapprove all flight plans, and is also responsible for:

a Providing preflight and inflight information to all pilots
b Keeping aircraft safely separated while operating in airspace controlled by ATC.

ATC FACILITIES Air Traffic Control consists of various communication facilities that provide voice communication between ATC controllers and pilots. The ATC facilities are divided into:

a Airport terminal facilities
b Air traffic services

AIRPORT TERMINAL FACILITIES Airport terminal facilities provide information to pilots and control aircraft on the ground and in the air near airports. These facilities, as shown in Fig. 1-1, consist of:

a Control Tower
b Ground Control
c Departure Control
d Approach Control
e Automatic Terminal Information Service (ATIS).

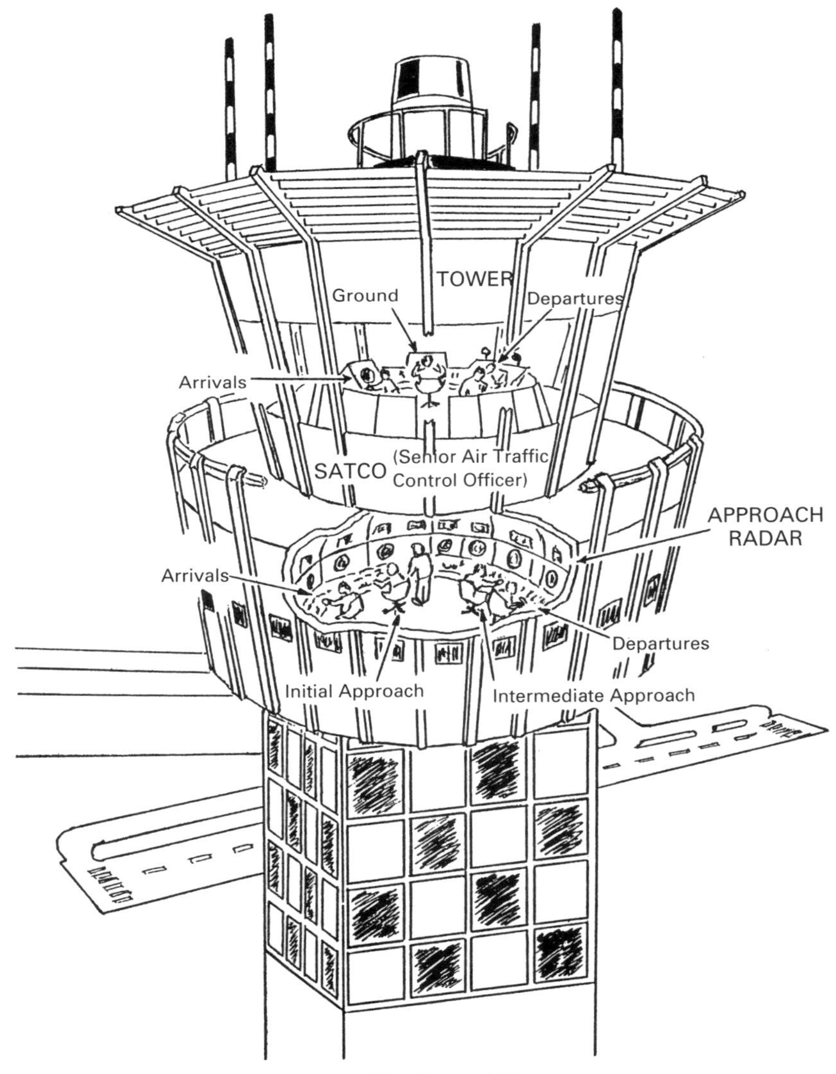

Fig 1–1 The Control Tower

CONTROL TOWER The control tower houses ATC personnel responsible for the control of aircraft on the ground and in the air in the immediate vicinity of an airport. These personnel use light signals, ground signals, radio communications and radar to control air traffic.

The ATC control tower has local control of air traffic, namely the air traffic in and around the traffic pattern and on the ground.

GROUND CONTROL Ground Control at an airport directs aircraft movement on the airport parking areas and taxiways. The ground controller is located in the airport tower, which allows an unobstructed view of the entire airport, and uses radio communications to:

a Provide precise taxying instructions to and from the take-off and landing runways
b Provide information on the ground hazards that exist along the taxying route
c Direct aircraft that have landed to the correct tie-down or service area.

DEPARTURE CONTROL Departure Control uses radio communications to give departure clearances and instructions to departing aircraft. This provides separation between departing and arriving aircraft to prevent collisions by:

a Using radar to ensure aircraft separation
b Informing pilots of the location of other aircraft in their vicinity
c Providing flight information to pilots (weather, hazards to flight, etc.).

APPROACH CONTROL Approach Control is responsible for keeping a safe distance (separation) between all aircraft arriving at an airport that are flying under Instrument Flight Rules (IFR).

Approach Control personnel use visual means and radar to maintain the separation, and they co-operate with the control tower personnel to ensure that local air traffic and IFR traffic maintain a safe separation. In small zones this unit also undertakes the function of the Zone Control Unit.

AUTOMATIC TERMINAL INFORMATION SERVICE More commonly referred to as ATIS, this is a radio broadcast that is continuously transmitted during aerodrome hours of operation on VHF or VOR discrete frequencies. The ATIS broadcast is received by tuning in to the ATIS frequency listed in the Aeronautical Information Services Manual or in the let-down plate for the airport.

The items of ATIS information are as follows:

a Airport name and broadcast code letter
b Aerodrome QNH (height above sea level)
c Surface wind direction (M) and speed

d Temperature and dP (differential pressure)
e Landing and take-off runways in use
f Instrument approach in use
g Details of aerodrome surface state (e.g., snow)
h Details of unserviceabilities of navaids
i Pertinent Notices to Airmen (NOTAMS) and Airmen Advisories.

Aerodrome Flight Information Service

Ref: RAC 3-9-1-6

While the international airspace is covered by FIRs which are provided with a Flight Information Service, many aerodromes not fully covered by an ATC unit are provided with an Aerodrome Flight Information Service (AFIS). This AFIS is operated by an Aerodrome Flight Information Service Officer (AFISO), who is licensed by the Civil Aviation Authority (CAA) and is subject to inspection.

When in communication with the AFISO, and an AFIS is being offered, it is identified by the suffix 'INFORMATION'. The AFISO is responsible for:

a Assisting pilots in preventing collisions by issuing information to all aircraft flying in the Aerodrome Traffic Zone
b Issuing information and instructions to aircraft on the manoeuvring area and apron, on behalf of the aerodrome operator. These instructions and information are given to prevent collisions between aircraft and vehicles or obstacles
c Informing aircraft of the aerodrome facilities and condition
d Alerting the safety services
e Initiating action on overdue aircraft.

Visual Flight Rules (VFR)

OUTSIDE CONTROLLED AIRSPACE (RULE 26) Aircraft flying outside controlled airspace should be at least 1,500 m horizontally and 1,000 ft vertically clear of cloud. Above FL100 the aircraft should have an 8 km flight visibility, and below FL100 there should be a 5 km flight visibility. (See Fig. 1-2 on page 14.)

The only exception from the above requirements occurs when the aircraft is below 3,000 ft above mean sea level (AMSL) with a flight visibility of 5 km, clear of cloud and in sight of the surface. For aircraft other than helicopters, the flight visibility of 5 km may be reduced to at least 1,500 m if the aircraft indicated airspeed (IAS) is 140 kt or less.

A helicopter flying below 3,000 ft may fly at any considered reasonable speed with relation to the visibility encountered, while in sight of the surface and clear of cloud.

INSIDE CONTROLLED AIRSPACE (RULE 25/27) Controlled Airspace covers Classes A to E (inclusive) airspace, and VFR flying is not permitted in Class 'A' airspace.

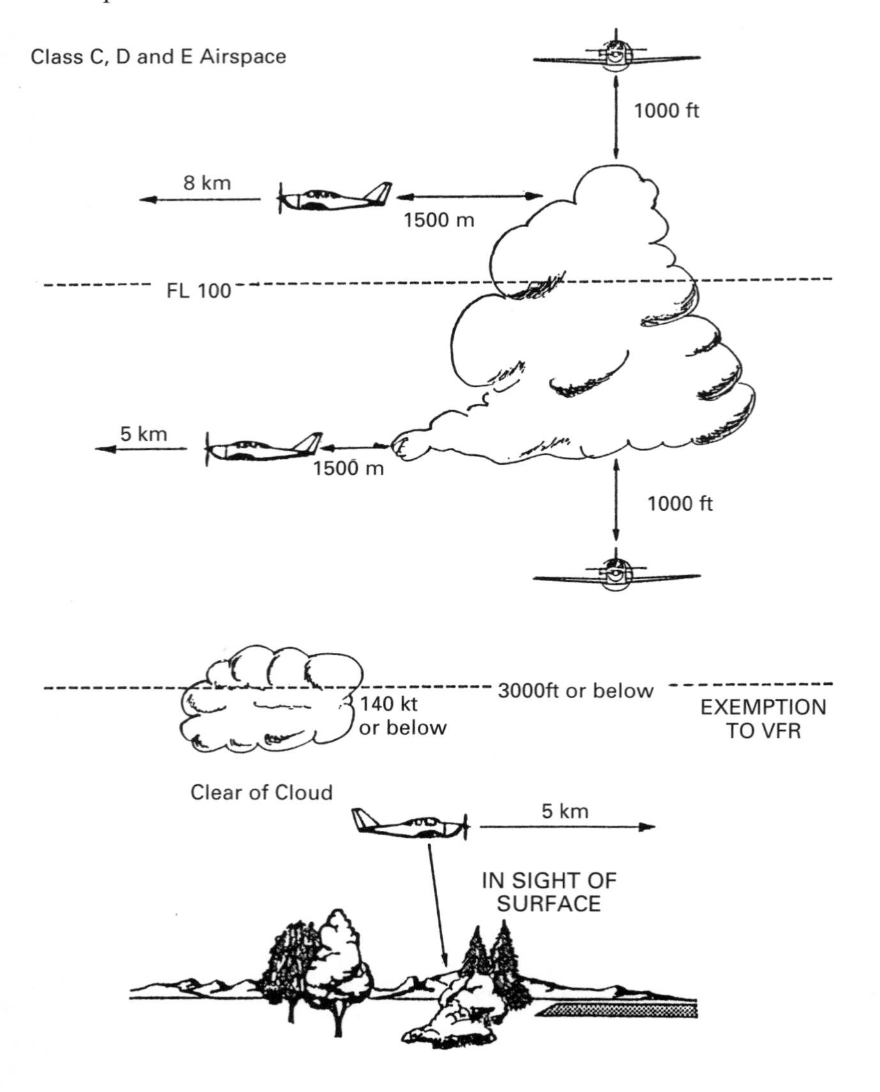

Fig 1–2 VFR Outside Controlled Airspace

CLASS 'B' AIRSPACE VFR flights in Class 'B' airspace should remain clear of cloud, and if above FL100 the flight visibility should be of at least 8 km, while below FL100 the flight visibility should be at least 5 km. (See Fig. 1-3)

CLASS 'C', 'D' AND 'E' AIRSPACE In these classes of airspace the flight visibility requirements are the same as for class 'B' above, being 8 km above FL100 and 5 km below FL100. The other requirements, both above and below FL100, are for at least 1,500 m horizontal and 1,000 ft vertical clearance of cloud. (See Fig. 1-2)

For Class 'B', 'C' and 'D' airspace a flight plan is required to be filed, and ATC clearance must be obtained before attempting to fly VFR. While flying VFR in this airspace the pilot is required to keep a continuous watch on the radiotelephone (RTF) and is to comply with all instructions given by the ATC unit.

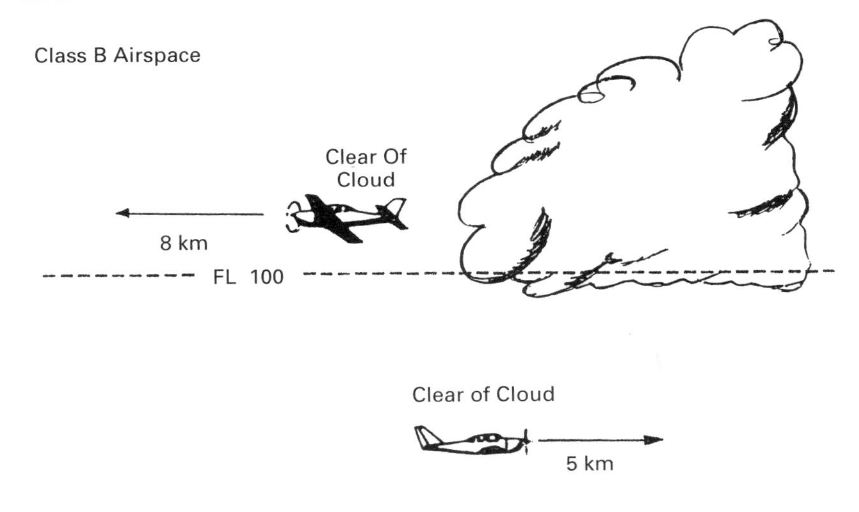

Fig 1–3 VFR in Controlled Airspace

The exceptions to the above requirements are:

1 Gliders flying during the day, which must have a flight visibility of at least 8 km and remain 1,500 m horizontally and 1,000 ft vertically clear of cloud.
2 Mechanically-driven aircraft without radio equipment whose commander has first obtained permission from the ATC unit. In this instance the aircraft must have at least 5 km visibility and remain 1,500 m horizontally and 1,000 ft vertically clear of cloud.

Instrument Flight Rules

RULE 22 Unless on a Special VFR flight, aircraft within a Control Zone must always fly IFR, and aircraft outside a Control Zone at night must also fly IFR.

RULE 29 For IFR flights in controlled airspace and outside controlled airspace, certain minimum height criteria must be satisfied. The criteria laid down in Rule 29 do not in any way overrule the criteria expressed in Rule 5 on low flying. However, so that aircraft are able to comply with IFR, all aircraft must not fly within 1,000 ft of any obstacle within 5 nm of the aircraft. The only exceptions to this rule are:

1 To allow an aircraft to land and take off
2 When authorised by the competent authority
3 When a special route has been notified
4 When the aircraft is in sight of the surface, clear of cloud and at an altitude of 3,000 ft or less.

OUTSIDE CONTROLLED AIRSPACE (RULE 30) When an aircraft is to be flown IFR outside controlled airspace, it is to be flown at a level that complies with either the Quadrantal Rule or the Semi-Circular Rule. This applies to all aircraft above transition altitude or in level flight above 3,000 ft AMSL, whichever is the higher. Each aircraft is flown at a level appropriate to its magnetic track and, within the UK, with a pressure setting of 1013.2 millibars (mb) set. The Quadrantal Rule applies to aircraft in airspace with a lower limit of 3,000 ft and up to all levels below 24,500 ft. The Semi-Circular Rule applies to all levels above 24,500 ft. As 24,500 ft is the dividing level between Quadrantal and Semi-Circular Rules, this is an unusable level.

QUADRANTAL RULES The Quadrantal Rule divides the 360° compass card into four segments, as seen in Fig. 1-4.

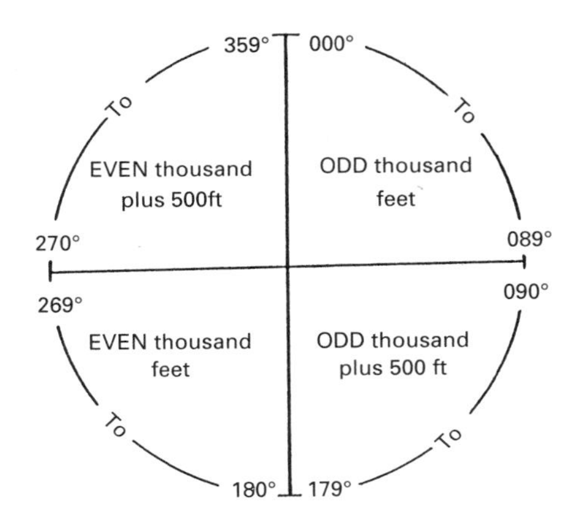

Fig 1–4 Quadrantal Rule

1 Aircraft flying on magnetic tracks from 000° to 089° must fly at odd thousands of feet (Flight Levels).
2 Aircraft flying on magnetic tracks from 090° to 179° must fly at odd thousands of feet (Flight Levels) plus 500 ft.
3 Aircraft flying on magnetic tracks from 180° to 269° must fly at even thousands of feet (or Flight Levels).
4 Aircraft flying on magnetic tracks from 270° to 359° must fly at even thousands of feet (or Flight Levels) plus 500 ft.

SEMI-CIRCULAR RULES The Semi-circular Rules divide the 360° compass card into two segments, as seen in Fig. 1-5.

1 Aircraft flying on magnetic tracks from 000° to 179° must fly at odd thousands of feet (or Flight Levels).
2 Aircraft flying on magnetic tracks from 180° to 359° must fly at even thousands of feet (or Flight Levels), up to and inclusive of FL280.

Above FL290 all cruising levels are in odd thousands of feet (or Flight Levels), but with a vertical separation of 2,000 ft on reciprocal tracks.

With reference to vertical separation, it should be noted that flights on reciprocal tracks have 1,000 ft separation (2,000 ft separation on same track) up to and including FL290. Above FL290 the separation on reciprocal tracks is 2,000 ft (4,000 ft separation on same track), to allow for the increase in instrument error on altimeters at higher levels. Because the changeover begins

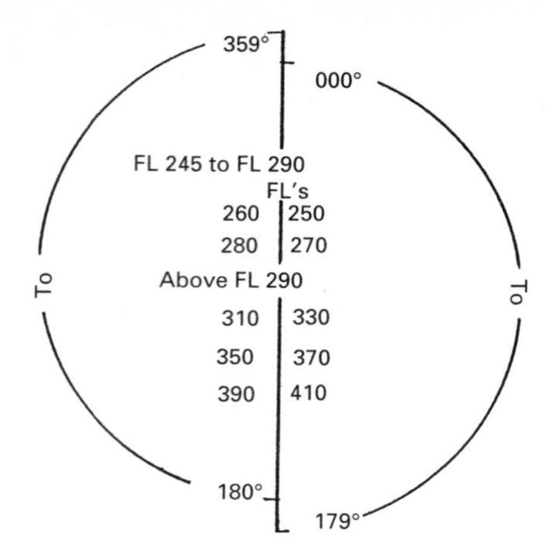

Fig 1–5 Semi–Circular Rule

at an odd-numbered flight level, all usable flight levels above FL290 are odd numbers. It will be noted on the semi-circular chart that the jump from FL280 is to FL310.

INSIDE CONTROLLED AIRSPACE IFR flight within controlled airspace requires a flight plan to be filed and ATC clearance to be obtained before an aircraft takes off from within controlled airspace. The filing of a flight plan gives the ATC service the intentions of the aircraft commander. It also enables the ATC service to monitor the progress of the flight, ensuring safe aircraft separation, and prepares the air traffic service in the event of a search and rescue being required.

All procedures, routes and flight levels specified in the flight plan should be adhered to unless further instructions are given by the ATC services.

The flight plan may be cancelled by the aircraft commander if he/she informs the ATC unit that he/she intends to continue to fly in VFR, ensuring that he/she is able to continue to fly in uninterrupted visual meteorological conditions (VMC) while remaining in controlled airspace. If the commander of the aircraft has to deviate from the provisions allowing the VFR flight to ensure the safety of the aircraft, then he/she should notify ATC as soon as possible.

When flying in controlled airspace and a landing is made or the aircraft leaves controlled airspace, ATCU must be informed unless they have otherwise cancelled the flight plan.

SPECIAL VFR *RAC 1-6 (12 Dec 91)* Occasions arise when a pilot who is unable to comply with the IFR wishes to transit, depart from or arrive within a control zone. In certain circumstances this can be accomplished by flying Special VFR. Special VFR is when the pilot has ATC authorisation and complies with ATC instructions within those control zones. This could be a flight at any time in a control zone which is classed as 'A' airspace, or in instrument meteorological conditions (IMC) or at night in any other control zone.

When permission has been granted for Special VFR the aircraft must remain clear of cloud and in sight of the surface, and the pilot must comply with any instruction given by that unit. Special VFR can be requested before take-off or once airborne, in which case all requests should be made 5–10 min before the specified estimated time of arrival (ETA) for the selected entry point. If the request is made from an aerodrome within a control zone, Special VFR clearance will only be issued if the visibility is greater than 600 ft.

Aircraft with an all-up-weight exceeding 5,700 kg and which are capable of flying under IFR may be granted a request to fly Special VFR, but this is only granted in exceptional circumstances.

Note **1** Flights requesting Special VFR must give the ATC details of the callsign, aircraft type and the pilot's intention. However, a full flight plan is not required.

2 Flights under Special VFR into London/Heathrow and also helicopters flying in the London Area are covered by special rules outlined in the *UK Air Pilot*.

RADIO FAILURE WHEN FLYING SPECIAL VFR If an aircraft is about to enter a control zone (CTR), having been given special VFR clearance, and a radio failure occurs, the aircraft should remain well clear and not enter the CTR.

In all cases of radio failure when a secondary surveillance radar (SSR) transponder is carried, the pilot should select Code 7600 and Modes 'A' and 'C'.

If the pilot is unsure as to whether the transmitter or the receiver is unserviceable (or both), or has reason to believe that the transmitter is serviceable, the pilot should transmit blind, stating condition and intentions and giving position reports.

When an aircraft is in the CTR when a radio failure occurs, if it is inbound to an aerodrome within the CTR it should proceed and land as soon as possible in accordance with the Special VFR clearance. Visual signals may be given from the ground when the aircraft is in the aerodrome traffic circuit.

If the aircraft is flying through the CTR under Special VFR when radio failure occurs, the pilot should leave by the most direct route while remaining at or below the cleared altitude. Weather limitations should not be exceeded, and areas of dense traffic and clearance of obstacles must be taken into consideration.

Air Traffic Services

ATC has the responsibility to maintain the separation between aircraft en route from one airport to another, and accomplishes this by providing the following air traffic services:

a Flight Information Service (FIS)
b Alerting Service
c Air Traffic Advisory Service.

FLIGHT INFORMATION SERVICE *(REF: RAC 3-8-1-1)* The FIS is an en route service provided by the ATCC through an FIR controller to aircraft flying outside Controlled Airspace and Advisory Routes. The FIS is for the purpose of communicating advice and information to pilots that is useful for the safe and efficient conduct of flights. The FIS includes:

a Providing en route and destination weather
b Providing en route communications with pilots on VFR flights
c Giving emergency assistance to lost IFR and VFR flights
d Relaying ATC clearances and position reports when requested
e Broadcasting aviation weather
f Performing initial search and rescue operations for aircraft
g Receiving and relaying pilot observed weather conditions.

FLIGHT INFORMATION CENTRE A Flight Information Centre (FIC) is a unit established to provide the FIS and Alerting Service.

ALERTING SERVICE The Alerting Service is provided by ATC to pilots flying outside of the area of an airport. It consists of:

a Notifying all the appropriate organisations that an aircraft is in a situation (emergency condition) that requires search and rescue aid
b Assisting all organisations in search and rescue operations as required.

AIR TRAFFIC ADVISORY SERVICE The Air Traffic Advisory Service (ATAS) is provided to pilots operating aircraft within an FIR to ensure that a safe separation is maintained between all IFR traffic. As a CONTROL service is a much more complete service than the advisory service, Advisory Routes and Airspace is only established within uncontrolled airspace.

ATC monitors and plots the present position of all IFR traffic within its area, and by communicating altitude and/or heading changes to an IFR aircraft, is able to maintain a safe separation.

SUMMARY ATC is responsible for ensuring that all aircraft are operated safely.

Airport terminal facilities control the operation of all aircraft on the ground and in the air in the vicinity of airports.

Air traffic services provide the FIS, Alerting Service and ATAS to aircraft operating outside airport terminal areas.

CHAPTER 2
DIVISION OF AIRSPACE

International Civil Aviation Organisation

With the expansion of aviation and air travel it became apparent that, to standardise flying regulations, a common unit or body was required to monitor and legislate requirements in the interests of air safety. This unit was affiliated with the United Nations as a specialised international body dealing with aviation matters. It is now known as the International Civil Aviation Organisation (ICAO), its present headquarters being in Montreal, Canada.

Member nations of ICAO comprise most of the countries of the world, with a few Eastern European exceptions that may be expected to join in the foreseeable future. These member nations subscribe to ICAO rules and procedures, with the exception of a few national differences in procedures which have been accepted. Most of the world is split into ICAO regions known as Flight Information Regions (FIRs).

Flight Information Regions

Flight Information Regions are, in the main, basically national in limits, but are mainly dependent on air traffic growth. To this degree, the UK is split into two FIRs to cope with the density of air traffic. These air traffic regions are further subdivided into upper and lower airspace. The lower airspace is known as the Flight Information Region, and is responsible for airspace up to FL245. The airspace above FL245 is known as the Upper Flight Information Region (UIR). It is dependent on the upper airspace traffic density whether or not one UIR may cover the same geographical area as two or three FIRs, giving it wider lateral limits, as follows:

FIR	UIR	Country
Paris Marseille Bordeaux	France	France
London Scottish	London Scottish	United Kingdom

The density of the airspace over the UK is such that there are two FIRs, London and Scottish, and two UIRs, London, based at West Drayton, and Scottish, based at Prestwick. The FIR/UIR boundaries are such that they extend or overlap with other national boundaries, and this is accounted for in bilateral agreements with the countries concerned. The FIRs/UIRs are centred at and controlled from their appropriate ATCCs.

Air Traffic Control Centres

An ATCC is established within each FIR/UIR to fulfil the following responsibilities.

a The control of aircraft in controlled airspace within its region
b The maintenance of up-to-date weather information within its region, together with pressure values, transition levels and other information required to ensure the safe and efficient operation of aircraft
c The provision of flight information, as follows:
  1 Information on unusual or dangerous weather
  2 Serviceability state of aerodromes and their equipment
  3 Serviceability of navigation aids
  4 Other information pertinent to the safety of aircraft
  5 Meteorological information
d To provide a signals centre for the receipt and dispatch of aeronautical information and air movements
e To provide an alerting service to notify the appropriate organisation regarding aircraft in need of assistance.

When transitting from one FIR to another, the aircraft crosses an FIR boundary. When crossing an FIR boundary the pilot has to provide both FIR ATCCs with the following information:

a Aircraft callsign
b Aircraft position
c Aircraft altitude or flight level
d Time of reaching the boundary

e Heading or planned route
f ETA at next reporting point.

The crossing of an FIR boundary is a mandatory reporting point, whether in controlled or uncontrolled airspace.

Controlled Airspace

ATC has the authority and responsibility to exercise control over air traffic to prevent collisions between aircraft and to provide a safe flow of the air traffic. However, ATC's authority and responsibility only applies to the airspace which is classified as controlled airspace.

These controlled airspaces exist usually at terminal areas (e.g., airports, aerodromes, etc.) where the close proximity of airfields creates a high traffic density. In addition, air traffic on main routes is fed through other controlled airspaces which serve to channel the traffic, therefore allowing a system of 'en route' separation to be used to reduce and prevent the risk of collision.

Controlled airspaces are divided into two basic types:

a Control Areas
b Control Zones

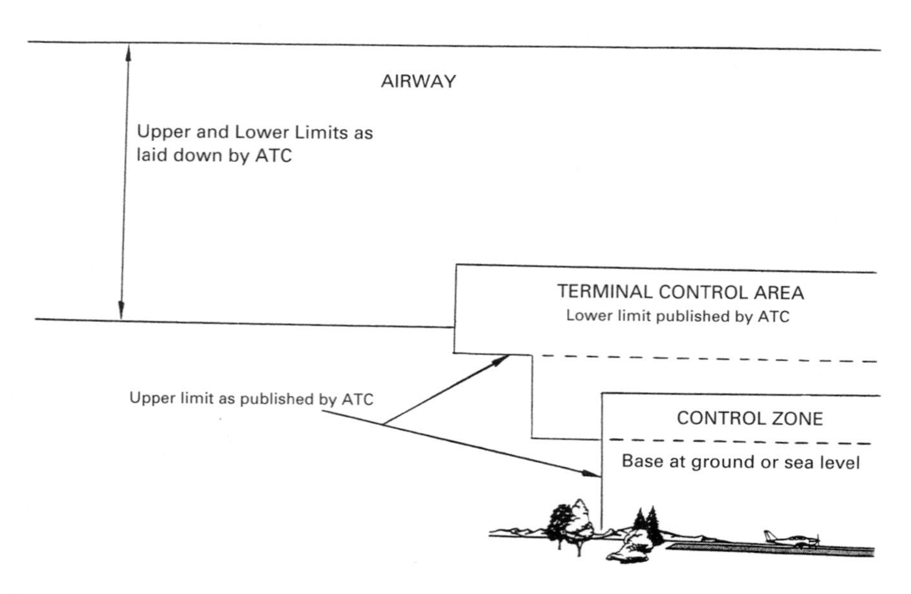

Fig 2–1 Controlled Airspace

Other descriptive names may be given to certain controlled airspaces to define their specific function; nevertheless, any controlled airspace remains essentially an 'area' or a 'zone'.

Controlled Airspaces and associated terms are defined as follows (and illustrated in Fig. 2-1):

a CONTROLLED AIRSPACE – an airspace of defined dimensions, within which air traffic control service is provided to IFR flights
b CONTROL AREA – a controlled airspace extending upwards from a specified height
c CONTROL ZONE – a controlled airspace extending upwards from the surface of the Earth
d AIRWAY – a control area or part of a control area, in the form of a corridor and marked by radio navigation beacons
e TERMINAL CONTROL AREA (TCA) – a portion of a control area normally situated where there is a confluence of airway traffic in the vicinity of one or more aerodromes.

CONTROL AREA Control Areas and TCAs have lower limits which are above ground level and upper limits as specified by national regulations. The area covered laterally varies considerably from area to area, but none are of a comparable size to the Oceanic Control Areas. Many control areas are split into sectors, and the lower limits of sectors may, and often do, vary.

CONTROL ZONES A Control Zone protects ALL aircraft flying within it, and extends from ground level up to a specified upper limit. This upper limit is normally the lower limit of a Control Area, although, if further control is necessary, the upper limit may be higher than the lower level of the Control Area. In lateral extent, Control Zones vary greatly from a two-mile-diameter circle to an irregular shape covering an area of several hundred square miles. One Control Area may be formed to cover several Control Zones.

A Control Zone is airspace within an FIR or Upper Flight Information Region (UIR), within which ATC service is provided to IFR flights. Control Zones permit ATC to:

a Control air traffic taking off and landing at airports when IFR conditions exist
b Restrict all air traffic to only IFR flights when the weather conditions are below VFR minima.

AIRWAYS An airway is defined as a control area or portion thereof established in the form of a corridor equipped with radio navigation aids. An airway is an air route with a centreline extending from one navigation aid or intersection to another navigation aid specified for that airway. Airways are shown on an aeronautical chart as having a blue centreline, and are normally referred to by a system of phonetics and numbers: e.g.: 'Alpha One', 'Golf Three', 'Romeo Six'.

AIRWAY DIMENSIONS The vertical dimensions of an airway normally begin at an altitude of 3,001 ft within the UK, where the lower limit is defined as a Flight Level or as specified on the chart. The upper limit is limited or unlimited, depending upon the ATC procedures used in the area of the airway, as illustrated in Fig. 2-2.

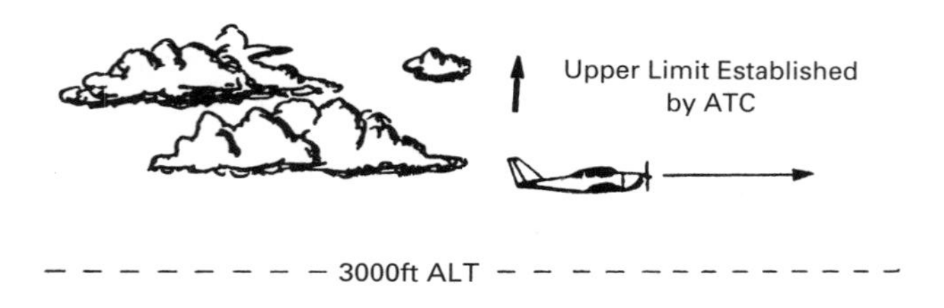

Fig. 2–2 Airway Vertical Limits

Each airway includes airspace within parallel boundary lines which mark the outer limit of the airway and is like a corridor in the sky. Each boundary line is 5 nm from the airway centreline, thereby making the total width of the airway 10 nm, as illustrated in Fig. 2-3.

TERMINAL CONTROL AREAS TCAs are also known as Terminal Manoeuvring Areas (TMAs), and are established to separate all traffic, both IFR and VFR, at the confluence of airways in the vicinity of large airports. A TCA extends upwards from the ground to an altitude designated by ATC to enable it to separate transit traffic from the traffic wishing to land and take off. A TCA is divided into ceilings and floors, as illustrated in Fig. 2-4, and allows aircraft to be routed to the zone boundary while other aircraft can be routed safely through the TCA on to another airway.

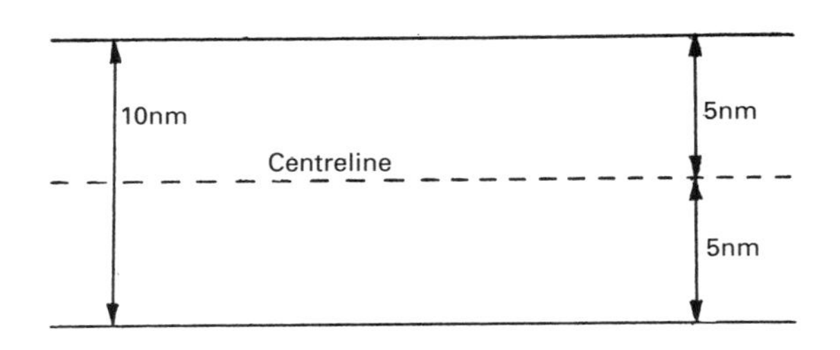

Fig 2–3 Airway Width Limits

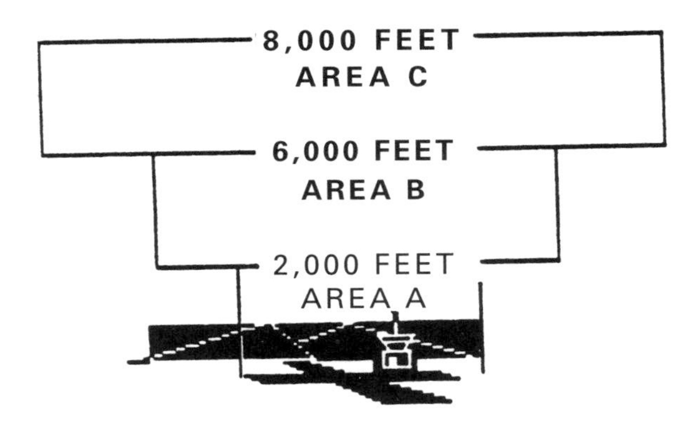

Fig 2–4 Terminal Control Area

ADVISORY AIRSPACE Advisory Airspaces are of two types, Advisory Area and Advisory Route.

An Advisory Area is a designated area within an FIR where air traffic advisory service is available.

An Advisory Route is a route within an FIR along which air traffic advisory service is available.

NOTE: ATC service provides a much more complete service than ATAS; Advisory Areas and Routes are therefore not established within controlled airspace. ATAS may be provided below and above controlled airspace. An Advisory Route (ADR or AD) is merely a recommended route in advisory airspace. No control will be offered, but 'advice' or 'suggestions' will be made to assist and allow the flight. The decisions normally made and required of an aircraft commander will, however, remain the responsibility of the aircraft commander. Advisory Airspace is mainly confined to Advisory Routes within the confines of the Classification of Airspace.

Chapter 3
Classification of Airspace

With effect from 14 November 1991, the airspace over the United Kingdom has been changed and categorised into seven different classifications from A to G inclusive. It is then split into IFR and VFR flights. The breakdown of the various classifications of airspace can be seen in the *UK Air Pilot* under the RAC section, on the chart RAC 3-0-1. A summary of the breakdown can be considered thus:

Class 'A' Airspace

This is controlled airspace in which no VFR flights are permitted. In this instance it is required that a flight plan must be submitted, ATC clearance is required for all flights, and full ATC service is provided. Separation is given to all aircraft and, to enable this to occur, radio communication is an essential requirement.

As this type of airspace covers Control Areas,which means airways, there is no speed limitation applied to aircraft in this category. Aircraft can therefore fly at different speeds at different flight levels without affecting the separation requirements.

Although the lower level of an airway is normally 3,000 ft above ground level (AGL), if the lower level is defined as a flight level, an absolute minimum altitude of 3,001 ft applies unless otherwise stated.

The lower levels of the airway take the terrain into consideration, and therefore certain minima are required for safe flying. The base of an airway must be at least 1,000 ft above any fixed obstacle within 15 nm of the airway centreline. As flight levels are flown in airways, which means that with the altimeter set at 1013.2 mb the actual height will vary, certain minima have to be applied. Under normal conditions an extra allowance of at least 500 ft is added to the original 1,000 ft stated above, giving at least 1,500 ft clearance. Where the actual base of an airway is defined as a flight level, then the absolute minimum clearance provided must not be less than 1,500 ft within 15 nm of any position of the airway centreline.

(RAC 3-2-1-2) 14 Nov 91

Class 'B' Airspace

This is controlled airspace covering all the UK, and includes all the Upper Air Routes, which are above FL245, and also the Hebrides Upper Control Area.

All of the IFR requirements are the same as for Class 'A' requirements in terms of the following:

a A flight plan must be filed
b ATC permission must be obtained before the Area is entered
c A continuous RTF watch must be kept on the appropriate frequency
d The flight must be conducted in accordance with ATC instructions
e Separation is given to all aircraft.

Class 'B' airspace applies to all IFR and VFR flights above FL245, and cruising levels are normally used in accordance with the semi-circular rules. These cruising level allocations may vary when ATC is matching the levels with those used for oceanic airspace crossings.

Aircraft on eastbound or westbound flights through the Scottish UIR should check with the *UK Air Pilot* for the correct procedures, as the flight planning must be correctly co-ordinated to match up with the oceanic air routes.

Although VMC minima refer to airspace between FL100 and the surface, this specification is currently left open for any future change in legislation, and therefore corresponds to VFR flights in the class 'C' to 'E' airspace. Therefore, under current legislation VMC minima is restricted to 8 km visibility and clear of cloud.

A Military Mandatory Radar Service Area (MRSA) covers a large portion of the Upper Airspace Control Area (CTA). All military aircraft flying between FL245 and FL660 within the MRSA operate under a radar control/procedural control service.

Military aircraft flying in the Upper Airspace may not necessarily be receiving an ATC service, owing to the nature of their individual tasks. This is more so over the North Sea area, and therefore civil aircraft operators should ensure that their flight plan structure follows the published ATS Route Structure. *(RAC 3-3-4, 12 Dec 91)*

VMC Minima for Class 'C', 'D' and 'E' Airspace

ABOVE FL100

1 8 km horizontal visibility
2 1,500 m horizontal clearance from cloud
3 1,000 ft vertical clearance from cloud.

BELOW FL100

1 5 km horizontal clearance from cloud
2 1,500 m horizontal clearance from cloud
3 1,000 ft vertical clearance from cloud.

The above criteria may affect VFR circuit flying at some aerodromes, and therefore short-term exemptions may be in force which will be stated on current Notices to Airmen (NOTAMS). To cover the long-term effect of this situation, amendments to the *UK Air Pilot* will be introduced as and when necessary.

Class 'C' Airspace

Under IFR conditions Class 'C' airspace allows for the separation of IFR traffic from other IFR traffic, and also from VFR traffic. It is also a requirement to have radio communication and ATC clearance. This enables a full ATC service to be available.

VFR conditions allow separation from IFR traffic from the ATC service, while VFR traffic information and traffic avoidance advice is available on request. Radio communication and ATC clearance is required, and a speed limitation of 250 kt IAS below FL100 is imposed.

NOTE: At the time of publication there is no Class 'C' airspace allocated within the UK.

Class 'D' Airspace

This is controlled airspace which gives IFR separation from other IFR traffic. This is accomplished through the ATC service, which will also provide VFR traffic information. Avoidance advice on VFR traffic is available on request. Other Requirements (IFR and VFR) are:

1 Radio communications
2 ATC clearance
3 A speed limitation of 250 kt IAS below FL100.

The ATC service will supply traffic information on all other flights when under VFR, although no VFR separation is provided.

Note 1: Alderney comes under the Guernsey legislation, which does not take account of the ICAO Airspace Classifications and associated VMC criteria at the present time.
Alderney has a Special Rules Zone which comes under the Class 'D' airspace for administrative purposes, and information pertaining to

Alderney can be found in the Class 'D' Controlled Airspace section of the *UK Air Pilot* RAC section.

Note 2: The Isle of Man has its own legislation and its own Special Rules Airspace, and information for the Isle of Man is found in the *UK Air Pilot* RAC section.

Class 'D' airspace can be seen on the UK Chart RAC 3-0-1 in the colour green and annotated as Class 'D' airspace.

Class 'E' Airspace

This is controlled airspace which is outlined in purple on the RAC 3-0-1 chart. The criteria are as follows:

IFR Flights

1 Separation from other IFR traffic
2 Speed limitations of 250 kt IAS below FL100
3 Radio communications required
4 ATC clearance required.

ATC services are provided and, where practical, VFR flight traffic information is also given.

VFR Flights

1 There is no provision for VFR aircraft separation
2 Speed limitation of 250 kt IAS below FL100.

VFR flight traffic information will be given where possible.

VMC Minima for Class 'F' and 'G' Airspace

For VFR flights in Classes F and G airspace, as illustrated in Fig. 3-1, the VMC minima are as follows:

Above FL100

1 8 km horizontal visibility
2 1,500 m horizontal clearance from cloud
3 1,000 ft vertical clearance from cloud.

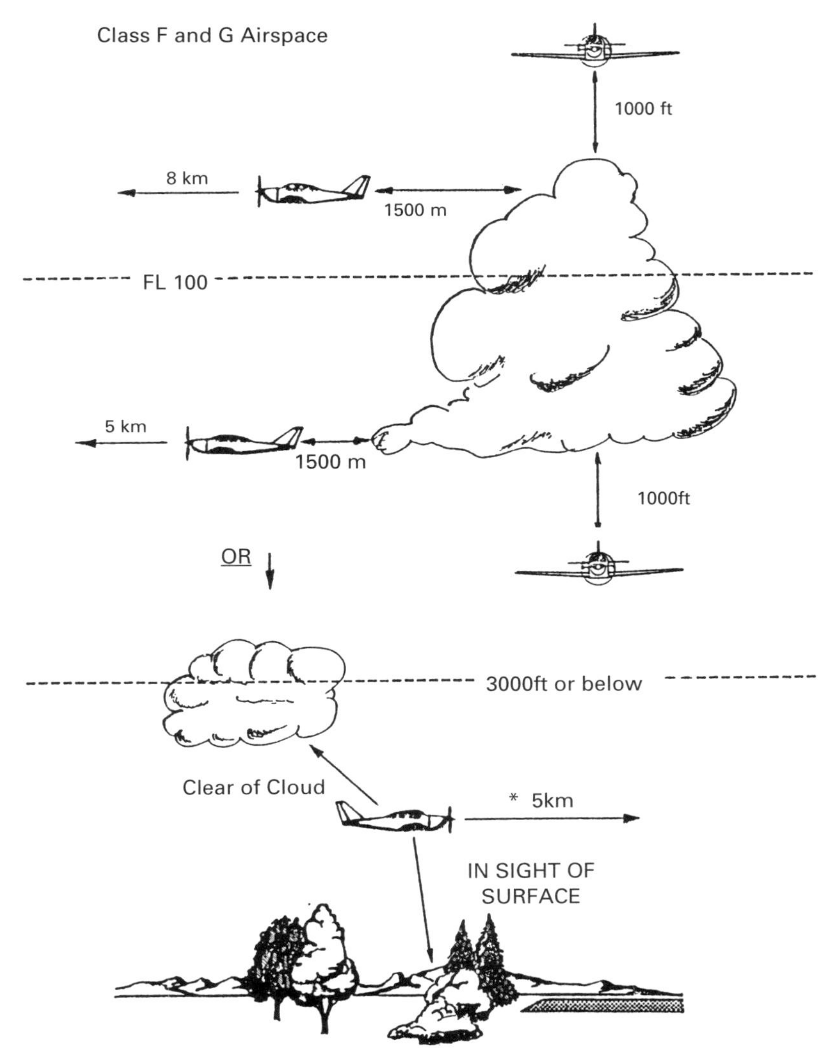

Fig 3–1 VMC Minima

BELOW FL100

1 5 km horizontal visibility
2 1,500 m horizontal clearance from cloud
3 1,000 ft vertical clearance from cloud.

BELOW 3,000 FT AMSL

1 Clear of cloud
2 5 km horizontal visibility unless below 140 kt, when flight visibility is allowed down to 1,500 m
3 In sight of the surface.

Class 'F' Airspace (RAC 3-7-1: 14 Nov 91)

This is uncontrolled airspace, but covers the Air Traffic Service Advisory Routes, and provides separation between participating IFR traffic. Pilots are urged to use this service to ensure safe separation between aircraft, and to ensure that the interests of safety are generally observed. As it is not a mandatory requirement, it is emphasised that all advice given to pilots of the participating aircraft is only in reference to the known traffic. As there may be unknown traffic in the vicinity, and military activity may be intense in some areas, it is the pilot's responsibility to ensure a safe look-out at all times. It is also the pilot's responsibility to ensure adequate terrain clearance, as some routes do not automatically provide 1,500 ft terrain clearance within 15 nm of track.

For aircraft wishing to participate in the Advisory Service, the following procedures must be adhered to:

1 A flight plan must be filed
2 When joining an ATS Advisory Route, approval must be sought not less than 10 min before entry point ETA when transitting from outside of controlled and Advisory Airspace
3 A listening watch must be maintained
4 Radio communication and navigation equipment must be carried when the ATS Advisory Route is directly associated with CAS
5 Quadrantal Rule level allocations are used.

IFR conditions otherwise allow for:

1 Separation of participating IFR traffic
2 ATAS and FIS
3 Speed limitation of 250 kt IAS below FL100.

Under VFR conditions separation is not provided, but an FIS is provided.

As these conditions meet the minimum ICAO standards they can be supplemented by Radar Advisory Service (RAS) or Radar Information Service (RIS), and pilots are urged to make use of these services whenever they are available.

Aircraft requiring to fly under IFR within Advisory Airspace shown in territorial procedures, but not electing to use the ATAS, are to pass the following information to the unit providing the service:

a Before entering the route or area – position, true airspeed, cruising level and intended route

b Within the route or area – any intended change in route or cruising level.

Class 'G' Airspace

This relates to all other airspace which is therefore primarily uncontrolled airspace and outside Advisory Routes, and under IFR and VFR conditions only an FIS is available. Separation is not provided, and a speed limitation of 250 kt IAS is imposed below FL100. The pilot is wholly responsible for separation and collision avoidance.

An RAS or an RIS may be available in Class 'G' airspace, depending on the controllers' workload.

To avoid confusion and to safeguard aircraft on airways, pilots flying in Class 'G' airspace should refrain from flying close to Controlled Airspace unless ATC clearance to enter has been given. Pilots should also refrain from flying parallel to airways for the same reason. However, where the lower limit of an airway is defined as a Flight Level, an aircraft may cross at an angle of 90° across the base of an en route airway without ATC clearance.

Military Aerodrome Traffic Zones (MATZ)

All military aerodromes have an Air Traffic Zone (ATZ) to cover all aircraft in the circuit taking off and landing. To provide increased protection to these ATZs by allowing a greater safety margin for aircraft on approach or climb-out, a MATZ is created around the ATZ. Civil flight through a MATZ does not require clearance, as recognition of them is not mandatory, but current Rules of the Air Regulations apply when penetration of the ATZ is required.

The normal dimensions of a MATZ comprise the airspace:

a Within 5 nm radius of the mid-point of the longest runway;

b From the surface to 3,000 ft above aerodrome level (AAL).

c Within an approach 'stub' which projects beyond the 5 nm radius limit to a further 5 nm in length, positioned along the runway centreline and 2 nm either side of the centreline (4 nm wide), from 1,000 ft AAL to 3,000 ft AAL.

When one MATZ is in close proximity to another, and especially when they overlap, the overall ATC will be given by one appointed Air Traffic Service (ATS) unit. The altimeter setting is on the MATZ QNH, but, when several MATZs are combined, the lowest value aerodrome QNH will be given, and it will be referred to as the 'Clutch QNH'.

Certain Royal Naval Air Stations still use QFE and therefore, when planning to penetrate any zones within this category, a check should be made by referring to the *UK Air Pilot* to ascertain prevailing conditions.

When penetrating a MATZ, an RAS will be given whenever possible. When this cannot be applied, however, a vertical separation of at least 500 ft may be used between known traffic.

As clearance to enter a MATZ is not mandatory, there may be unknown aircraft within the MATZ airspace. It is therefore the pilot's responsibility to maintain a careful look-out at all times, and also to maintain terrain clearance.

Before penetrating a MATZ, pilots should contact the MATZ controller when the aircraft is at least 5 min or 15 nm from the zone boundary, whichever is the greater. Procedures for penetration of a MATZ can be found in the *UK Air Pilot, RAC 3-8-2-1 (14 Nov 91)*.

Lower Airspace Radar Service (LARS)

Within 30 nm of each participating ATS unit, radar/radio cover is made available to all aircraft outside controlled airspace between the surface and FL95 inclusive. The hours of operation depend on each ATS unit; however if it is not classed as a 24-hour operation, flying may make the ATS unit operational outside the standard published hours. Pilots are therefore advised to attempt to make contact with the ATS unit by making three consecutive calls, after which it can be considered unavailable.

Controller's advice given under the RAS is designed to enable certain minimum separation criteria. These criteria are:

1 A minimum horizontal separation of 3 nm between all identified aircraft working the same unit

2 A minimum horizontal separation of 5 nm between identified aircraft and other observed aircraft, unless it is known that a 1,000 ft vertical separation exists.

It is not mandatory for any pilot to heed the advisory avoiding instructions. If the pilot decides not to, however, he/she should inform the controller immediately, and then the pilot must assume the responsibility for initiating subsequent avoiding action.

The controller must be warned before any changes of heading or flight level/altitude when under a radar service, and vertical separation should be in accordance with the Quadrantal Rule.

RAC 3-8-3-1 (14 Nov 91)

Military Middle Airspace Radar Service (MARS)

This service is similar to the LARS, but it covers airspace between FL100 and FL240, except for Brize Norton, which has an upper limit of FL150.

RAC 3-8-8-1 (14 Nov 91)

> **Caution**: Due to a Units capacity being exceeded in the LARS and MARS, controllers may not be aware of some aircraft, so pilots should maintain a careful look-out at all times.

Aerodrome Traffic Zones

Ref: *RAC 0-3*
RAC 3-9-2-1/2
ANO RULE 39

Aerodrome Traffic Zones (ATZs) are defined in the *UK Air Pilot* RAC 0-3 section. Briefly, the dimensions can be considered to be as follows:

LONGEST RUNWAY LENGTH OF 1,850 M OR LESS

1 From the surface to a height of 2,000 ft above the level of the aerodrome.
2 A circle of 2 nm radius centred on the midpoint of the longest runway.

LONGEST RUNWAY LENGTH OF MORE THAN 1,850 M

1 From the surface to a height of 2,000 ft above the level of the aerodrome.
2 A circle of 2.5 nm radius centred on the midpoint of the longest runway.

If the runway length is 1,850 m or less but the boundary of the zone would leave less than 1.5 nm from the end of the runway, that aerodrome's radius would be considered as for runways greater than 1,850 m.

No aircraft is allowed to fly, take off or land within an ATZ without prior permission, unless there are ATZs overlapping and the other aerodrome is the controlling aerodrome. In this case permission is required from the controlling aerodrome.

The ATZs are only valid during their open hours of watch, when obeyance of the flight rules and instructions are compulsory. Pilots are required to maintain a continuous RTF watch in an active ATZ, and to relay their position and height to the controller on entering the zone and immediately before leaving it.

ATZs are located within other classes of airspace, and the overriding requirements are either that of the type of airspace or Rule 39 of the ANO Rules of the Air Regulations, whichever is the more stringent.

Reserved or Special Use Areas

Reserved or special use areas consist of airspace in which the operation of aircraft is limited or not permitted. These areas are shown on aeronautical charts by a blue box with slash marks. The type of area is shown by a letter in the box, and its number designates its reference, the first number relating to its latitude. The special use areas and letters are:

PROHIBITED AREAS	P
RESTRICTED AREAS	R
DANGER AREAS	D.

A nationality designator, EG, precedes the special use area initial for the United Kingdom, though it is omitted from chart RAC 5-0-1 for reasons of clarity. The RAC 5-1 section covers all the types of special use areas in detail. However, the Chart of UK Airspace Restrictions at RAC 5-0-1 uses the following chart symbols with examples which are colour coded:

Prohibited Area (Purple)	P047/2.2
Restricted Area (Purple)	R312/2.1
Danger Area (Red)	
Scheduled	D127/12
Activated by NOTAM (Red)	D055/55
Military Training Area (Yellow)	LINCOLNSHIRE MTA FL245–FL350
Bird Sanctuary (Green)	MINSMERE/2
High-Intensity Radio Transmission Area (Blue)	FYLINGDALE/31.5
Air-to-Air Refuelling Area (Black)	AREA 1 FL220–FL280

PROHIBITED AREA A prohibited area is a designated part of the airspace of defined dimensions in which the flight of an aircraft is NOT PERMITTED.

RESTRICTED AREA A restricted area is an airspace of defined dimensions within which the flight of an aeroplane is not prohibited, but is subject to certain limitations. Restricted areas denote the existence of hazards to flight, such as artillery firing, military flight training, aerial gunnery, or the flight of guided missiles. Entrance into a restricted area is allowed only after obtaining clearance from the authority that controls the restricted area. Restricted areas are labelled EGR on aeronautical charts. Flight of aircraft is restricted in accordance with specified conditions.

Danger Area A danger area is airspace of defined dimensions (certain size) within which activities dangerous to the flight of aircraft may take place or exist at such times as may be notified.

The airspace above ammunition dumps or an oilfield can be classified as a danger area owing to the possibility of an explosion occurring. Danger areas are labelled EGD on aeronautical charts.

Special Rules Zone Special Rules Zones are defined areas of airspace around certain airfields, and have a radius of 5 nm around the airfield and extend up to 2,000 ft. A pilot who intends to fly within this area must:

a Obtain permission of the ATC unit at that aerodrome at least 10 min before ETA at the zone boundary
b Maintain a continuous listening watch on the appropriate frequency.

These rules also apply at some airfields whose areas do not conform with the above.

Special Rules Zones currently apply only to the Channel Islands and the Isle of Man.

CHAPTER 4
RULES OF THE AIR

Introduction

Aircraft in and out of controlled airspace must fly in such a manner as to avoid accidents and allow safe separation between conflicting traffic. To enable the safe separation and avoidance of collisions to take place, a 'highway code' of the air has to be adhered to, and this is known as the *RULES OF THE AIR 1991*. These rules are laid down in the Air Navigation Order of 1989 (ANO), and are very explicit in describing the 'do's and dont's' for the pilot.

The rules apply to aircraft at all times in the air and on the ground. When an aircraft is in controlled airspace, its pilot is still responsible for ensuring that it does not collide with any other aircraft. This means that if, in obeying the rules, a collision will occur, he must take all the necessary steps to avoid that collision, even if it means diverting from the rules. In such a case he should make a report in writing to the CAA, or the appropriate authority, within ten days of the occurrence.

To ensure that collisions do not occur, aircraft should not be flown close to each other whereby they may endanger each other. This general rule also applies to formation flying or attempted formation flying. Formation flying may be approved only if all the aircraft commanders have agreed to do so. Under the rules, some aircraft have the right of way and some must give way. As no signals or signs can be constructed in the sky, it is assumed that all pilots will adhere to these rules and give way when necessary. When an aircraft has to give way, it should adhere to the rules. However, unless it is passing well clear of the other aircraft, the yielding aircraft should avoid crossing ahead of the other aircraft and also avoid passing over or under it. The aircraft that has right of way should always maintain its course and speed unless it is avoiding a collision.

If an aircraft is towing a glider it must be considered to be a single aircraft, and its total length must not exceed 150 m. Overall command of the combination is the responsibility of the commander of the towing aircraft.

Types of Aircraft

Although the rules of the air are universal and cover all types of aircraft, certain conditions are controlled by the types of aircraft. Some aircraft are governed purely by the weather conditions (i.e. winds, air pressures and currents, etc.), while others have various types of controls and are able to manoeuvre to avoid collisions. Among the aircraft with controls, there is again a difference between aircraft as to the amount of control available. The manoeuvrability and control of all types of aircraft are therefore not always the same, and the more manoeuvrable aircraft must give way to the other aircraft. This is summarised as follows:

1 Flying machines must give way to airships, gliders and balloons
2 Airships must give way to gliders and balloons
3 Gliders must give way to balloons.

Aircraft controllability can be further classified as follows:

A POWERED AIRCRAFT – FULLY CONTROLLABLE This type is an aircraft with engines and full flying controls. Its pilot is able to control its attitude about all its three axes, and is able to control its speed up to its maximum allowable, and is therefore able to control its rate of climb and descent (Fig. 4-1).

Fig 4–1 Powered Aircraft
(Fully Controllable)

B POWERED AIRCRAFT – RESTRICTED CONTROL This type of aircraft is manoeuvrable, but is slow in its operation. Its manoeuvrability in the climb and descent is also slow, and therefore it is restricted overall. Airships come into this category (Fig. 4-2).

Airship

Fig 4–2 Powered Aircraft Restricted Control

C NON-POWERED AIRCRAFT – MANOEUVRABLE CONTROL This type of aircraft is normally a fixed-wing aircraft, called a glider, which has flight controls like those of a powered, fixed-wing aeroplane but does not necessarily have an engine. However, this type of aircraft is launched either by a ground winch or towed by another aircraft. The glider remains airborne by the skill of the pilot and the thermal air currents available (Fig. 4-3).

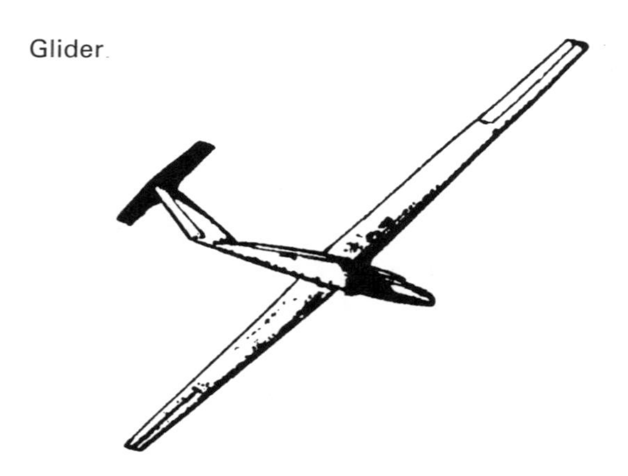

Fig 4–3 Non-Powered Aircraft Manoeuvrable Control

D NON-POWERED AIRCRAFT – LIMITED CONTROL Although kites come into this category, it is normally manned balloons that are referred to when considering this type of aircraft. They have very limited control in the ascent and descent, but directional control is governed totally by air currents and the vessel is therefore not manoeuvrable at all (Fig. 4-4).

Fig 4–4 Non-Powered Aircraft Limited Control

This classification can be summarised in Table 4-1 as follows:

Aircraft			
Lighter-than-air		Heavier-than-air	
Non-power-driven	Power-driven	Non-power-driven	Power-driven (flying machines)
Free balloon	Airship	Glider	Aeroplane (landplane)
Captive balloon		Kite	Aeroplane (seaplane)
			Aeroplane (amphibian)
			Aeroplane (self-launching motor-glider)
			Powered lift (tilt-rotor)
			Rotorcraft: Helicopter, Gyroplane

Table 4-1

It is up to all pilots to know and understand the different types of aircraft and their manoeuvrability, so that when another type of aircraft is encountered, the pilot can take the correct action to avoid a collision.

Converging Aircraft (Fig. 4-5)

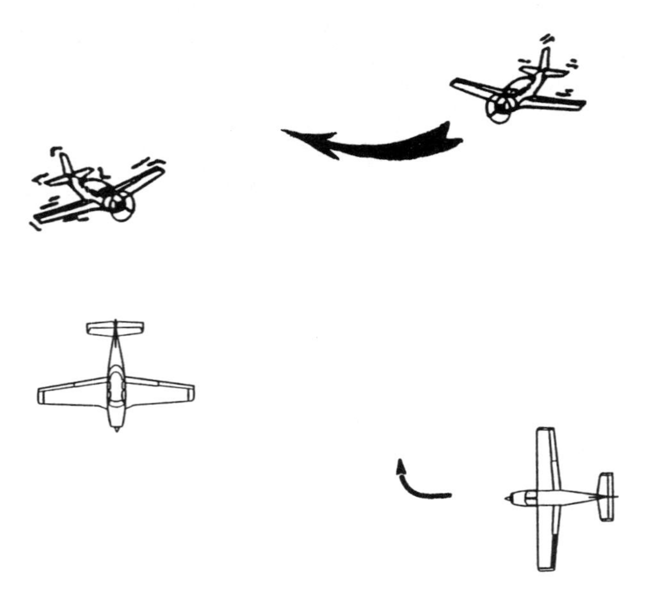

Fig 4–5 Converging Aircraft

When two aircraft are converging at approximately the same altitude, the aircraft on the right has the right of way (the other aircraft must manoeuvre to avoid a collision). This rule applies subject to the above conditions, and also to the rule that power-driven aircraft must give way to aircraft towing another aircraft or object.

Aircraft Approaching Head-on (Fig. 4-6)

When two power-driven aircraft are approaching head-on and there is a danger of a collision:

BOTH AIRCRAFT MUST TURN TO THE RIGHT TO AVOID A COLLISION.

This rule applies also in the case of a power-driven flying machine and an airship.

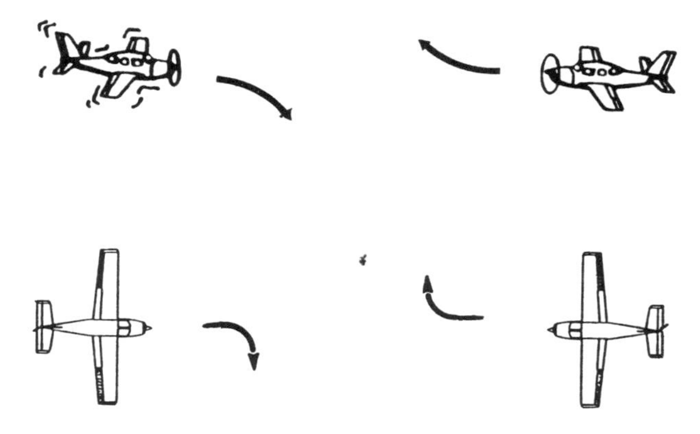

Fig 4–6 Approaching Head-On

Overtaking (Fig. 4-7)

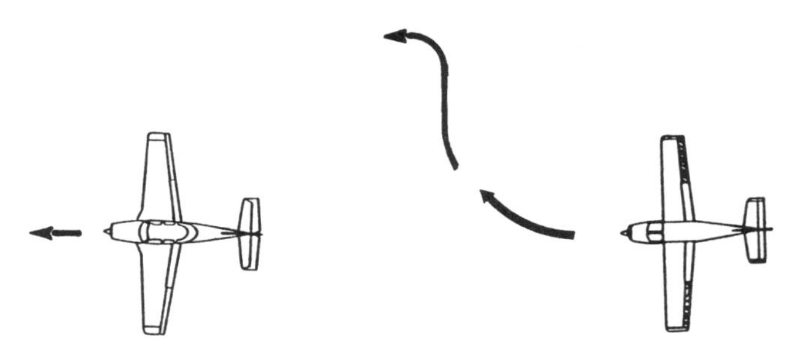

Fig 4–7 Overtaking

An overtaking aircraft is an aircraft that is approaching another aircraft from the rear. When an aircraft overtakes another, the aircraft being overtaken has the right of way. The overtaking aircraft, whether climbing, descending, or in level flight, must alter its course to the RIGHT to ensure that it keeps well clear of the aircraft being overtaken. Provided that a glider overtaking another glider in the UK may alter its course to the right or to the left.

Priority of Landing Aircraft

An aircraft that is landing or in the final phase of landing (on final approach) has priority over all other aircraft except an aircraft that has an emergency. Other aircraft flying in the area or on the ground must give way to the landing aircraft and not interfere in any way with the landing aircraft.

When two aircraft are approaching an airport for a landing, the aircraft at the lowest altitude has the landing priority. However, if the aircraft at the higher altitude is on the final approach for a landing, it has priority. The lower aircraft must not cut in front of or overtake the aircraft on final approach. The overtaking aircraft must take whatever action necessary to avoid the other aircraft, such as starting a climb (Fig. 4-8).

Fig 4–8 Priority on Landing

These rules apply except when the ATC unit has informed any aircraft of a different priority for landing.

Priority in Emergencies (Fig. 4-9)

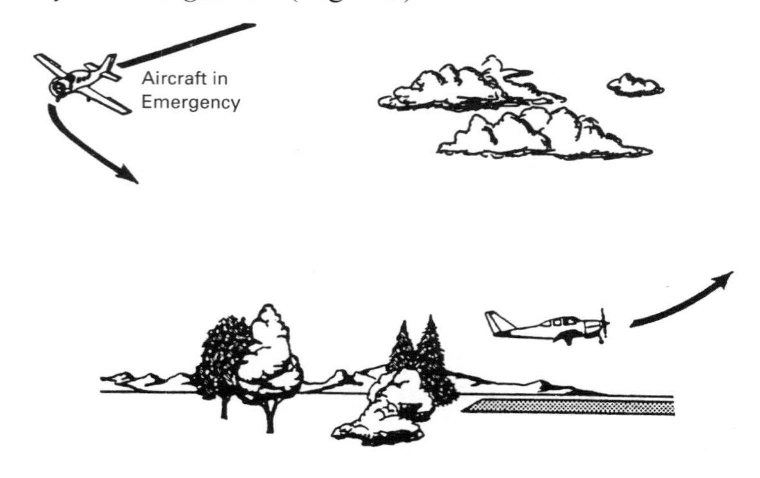

Fig 4–9 Landing Priority in Emergency

When an aircraft has an emergency situation and must land immediately, ATC will inform all aircraft in the local area. In this situation, the aircraft with the emergency has priority over all other aircraft. Any aircraft on final approach or on the active runway must take whatever action necessary to permit the aircraft with the emergency to land (Fig. 4–9).

Standard Traffic Patterns

All the above rules apply to prevent an aerial collision, and therefore traffic patterns should be observed to increase the safety factor. To achieve this, all aircraft should conform to a standard traffic pattern, which is a left-hand circuit. Any variance from the standard traffic pattern is indicated in the SIGNALS SQUARE on the aerodrome, otherwise the standard rule applies.

Airspeed Limitations (Fig. 4-10)

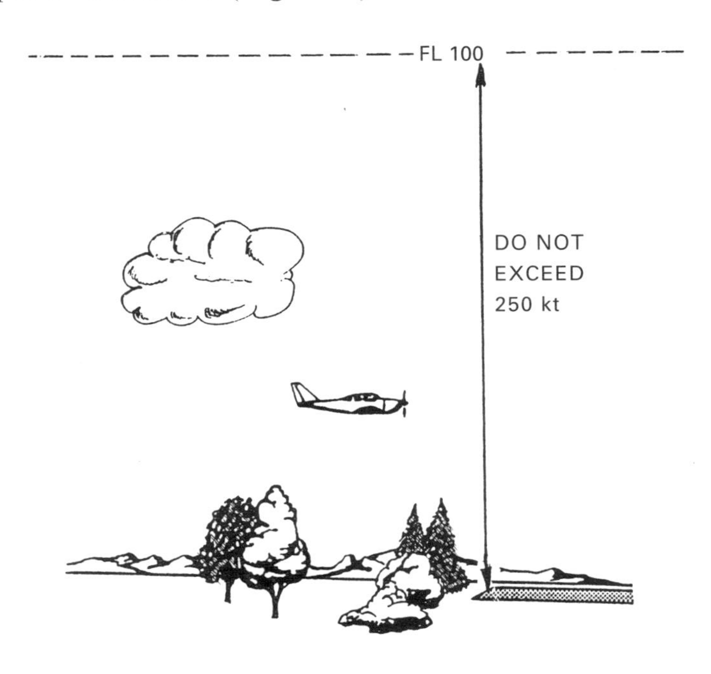

Fig 4–10 Airspeed Limitation

An airspeed limitation applies to all aircraft when flying below FL100. This states that aircraft must not exceed 250 kt when flying below FL100 unless the pilot has written permission from the CAA. This requirement does not apply to the following:

a Flight in Class 'A' airspace
b VFR or IFR flight in Class 'B' airspace
c IFR flight in Class 'C' airspace or VFR flight in Class 'C' airspace when authorised by the appropriate ATCU
d IFR or VFR flight in Class 'D' airspace when authorised by the appropriate ATCU
e Flights for authorised exhibition conditions or flights covered under Schedule 2 of the ANO.

Aerobatic Flight

Aerobatic flight manoeuvres must not be conducted unless the placard (limitations plate) in the aircraft cockpit indicates that the relevant manoeuvres are approved for the aircraft. This information should also be checked with the aircraft's flight manual before take-off. When conducting aerobatic manoeuvres you must not:

a Fly over congested areas (cities, towns or settlements)
b Fly in controlled airspace unless special clearance is obtained from ATC (i.e. air displays, shows, etc.).

Rights-of-Way on the Ground

Take-off, taxying and landing priorities are established by ATC to ensure that accidents do not occur between aircraft taxying, taking off and landing. These priorities also prevent one aircraft from interfering with another during the critical stages of a landing. During ground operations (taxying), right-of-way rules prevent collisions.

Right-of-Way During Taxying

When aircraft are taxying on an airport parking area (tarmac) and taxiways, the following rules must be adhered to:

1. A taxying aircraft must give way to all aircraft taking off and landing (this also applies to helicopters taking off from heliports in the vicinity of the parking and taxi areas).
2. When two taxying aircraft approach head-on, both must stop and then alter their courses to the right to ensure clearance between the aircraft.
3. When two taxying aircraft are on a converging course, the aircraft on the right has the right-of-way and the aircraft on the left must stop or turn to avoid a collision.

4 When a taxying aircraft overtakes another taxying aircraft, the overtaken aircraft has the right-of-way and the overtaking aircraft must remain well clear of the overtaken aircraft, and any change of course should be made to the left.

When an aircraft is taxying out for take-off, it must not interfere with aircraft taking off and landing. A taxying aircraft must stop far enough short of the active runway so that no part of the taxying aircraft extends over the runway. At some airports, a hold line is painted over the taxiway, and no part of the aircraft must extend past this line.

There are also set rules for vehicles and aircraft when they are approaching each other. Vehicles must give way to aircraft. However, if a vehicle is towing another aircraft, the aircraft must give way to the combination of the vehicle and the towed aircraft. This can be summarised in the order or priority, which is:

a Aircraft landing and taking off
b Vehicles towing aircraft
c Aircraft taxying
d Vehicles.

Members of the public and their private vehicles are not allowed on to any active part of the aerodrome. This rule applies except where they have the permission of the person in charge of the aerodrome, or where an ATC unit is operating, when the permission of the ATC unit is required. A further exception to this rule is where a public right-of-way exists.

Aircraft should not move on the manoeuvring area of the aerodrome or the apron without the permission of the ATC unit or the person in charge of the aerodrome. This allows greater control of all the movements on the ground, and is designed to enable safe movement of all aircraft, vehicles and persons. Although these rules are designed for safe ground operation, they do not in any way overrule the pilot's responsibility in avoiding any obstacle which may cause a collision. Ground marshallers are available to assist the pilot in preventing a collision, but if, in the opinion of the pilot, a collision is possible, the pilot should take all possible steps to prevent it.

Right-hand Traffic Rule

To further avoid collisions, when an aircraft is flying in sight of the ground following landmarks such as roads, canals, railways, etc., the aircraft should fly to the right side of the landmark. This rule does not apply to aircraft flying in controlled airspace under ATC instructions.

Lights to be Shown by Aircraft

Flying Machines

ANTI-COLLISION LIGHTS Anti-collision lights are to be displayed day or night by aircraft fitted with them. An anti-collision light is a flashing white or flashing red light in the case of fixed-wing aircraft, and a flashing red light in the case of rotorcraft.

It must be displayed at all times on the apron when the aircraft is stationary if the engines are running, and at all times in flight.

If the anti-collision light fails during the day, the aircraft may continue its flight provided the light is repaired as soon as possible. If the light fails at night before take-off, the departure should be terminated if the light cannot be repaired or replaced immediately. If the light fails at night in flight, the aircraft should be landed as soon as safely possible unless the ATC has cleared the flight.

Helicopters operating from offshore installations are allowed to switch off the anti-collision light when stationary, to indicate to the ground personnel that it is safe to embark and disembark.

NAVIGATION LIGHTS These consist of three lights situated at strategic positions around the aircraft, to enable at least one light to be visible at all times. These three lights comprise:

- **a** A green light on the starboard side showing from dead ahead in the horizontal plane and through an arc of 110°. Its brilliance should be of at least five candela.
- **b** A red light on the port side showing from dead ahead in the horizontal plane and through an arc of 110°. Its brilliance should be of at least five candela.
- **c** A white light showing from dead astern to an arc 70° each side of the dead astern position (this makes a total angle of 140°) in the horizontal plane. The brilliance of the white light should be of at least three candela.

These three lights are illustrated in Fig. 4-11.

All of these navigation lights must also be visible through the vertical plane through angles of 90° above and 90° below the horizontal position. In the horizontal and vertical planes the angles stated should not be exceeded where possible. The angles of display are illustrated in the accompanying diagram. If the navigation lights fail before take-off, departure should be terminated if the lights cannot be repaired or replaced, and if the aircraft is in flight at the time of failure, it should land as soon as safely possible or as directed by ATC.

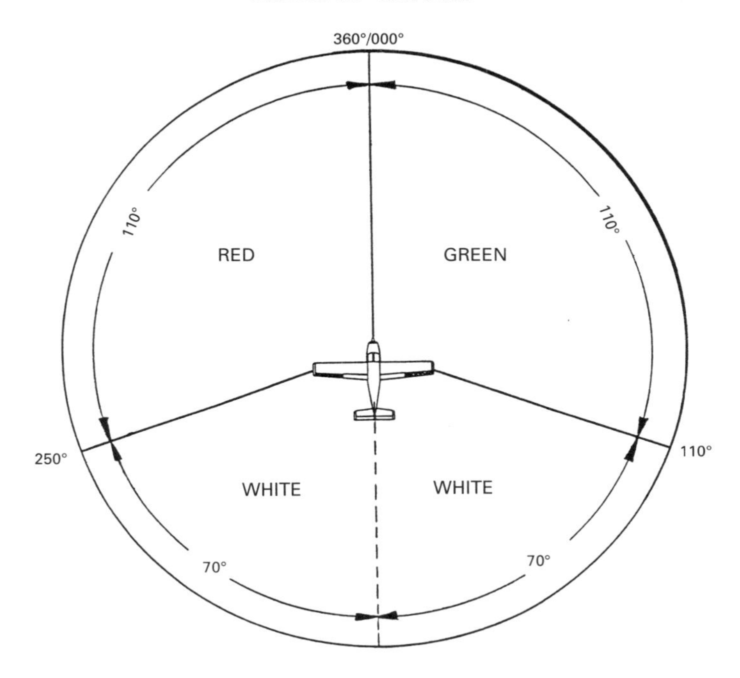

Fig 4–11 Navigation Lights Coverage

Aircraft Light Display Requirements

The conditions laying down the requirements for aircraft to display lights are explained in the following paragraphs.

1 A flying machine of maximum take-off weight unloaded (MTWU) exceeding 5,700 kg, and registered in the UK where the type certificate was issued on or after 1st April 1988, must display the following lights:

- **a** steady navigation lights, and
- **b** an anti-collision light.

2 A flying machine with a MTWU of 5,700 kg or less, and registered in the UK with a type certificate issued before 1st April 1988, must display:

- **a** steady navigation lights only, **OR**
- **b** steady navigation lights and an anti-collision light, **OR**
- **c** flashing navigation lights, flashing in alternation with a flashing white light of at least 20 candela, showing in all directions.

3 all other flying machines must display one of the following light conditions:

- **a** steady navigation lights only, **OR**
- **b** steady navigation lights and an anti-collision light, **OR**
- **c** flashing navigation lights, flashing together, **OR**
- **d** flashing navigation lights, flashing in alternation with a flashing white light of at least 20 candela, showing in all directions, **OR**
- **e** flashing navigation lights, flashing in alternation with a flashing red light of at least 20 candela, showing through angles of 70° from dead astern in the horizontal plane, **OR**
- **f** flashing navigation lights, flashing in alternation with both a flashing white light of at least 20 candela showing in all directions and a flashing red light of at least 20 candela showing through angles of 70° from dead astern in the horizontal plane.

In the event that the flashing lights may affect any member of the flight crew or dazzle an outside observer unreasonably, the commander may reduce the intensity or switch off the flashing light.

Free Balloons

A steady red light of at least five candela should be displayed at night by a balloon, and it should be suspended between 5 m (minimum) and 10 m (maximum) below the basket, or below the lowest part of the balloon if no basket exists.

Gliders

At night, gliders may display a steady red light or the same combination of lights afforded to any other flying machine.

Captive Balloons and Kites

If a height of 60 m AGL is exceeded by a captive balloon or kite being flown at night, it must display the following lights:

- **a** The mooring object to which the balloon or kite is moored must be set within a triangle of lights, the triangle being equilateral and all three sides being approximately 25 m each in length. The horizontal projection of the mooring cable must be aligned with one of the sides at an angle of 90°, this side being marked by two red lights. A third light, green in colour, enables the mooring object to be enclosed within the triangle. All three lights are to be flashing lights.
- **b** Two lights, each of 5 candela, one white and one red and showing in all directions, must be placed below the basket, or below the lowest part of

the balloon or kite. The white light should be not less than 5 m and not more than 10 m below the balloon or kite, and the red light should be placed 4 m below the white light.

c Groups of two lights of the same colour and power must be displayed along the mooring cable at intervals of not more than 300 m, measured from the red light mentioned above. If required, an extra set of lights must be displayed to ensure that a set of lights are visible below any cloud base.

If a balloon or kite is being flown by day at a height exceeding 60 m, streamers are to be attached to the mooring cable at distances not to exceed 200 m in the case of balloons and not to exceed 100 m in the case of kites.

Airships

When an airship is flying at night it is required to display navigation and anti-collision lights as for a flying machine, but they must all be of 5 candela. In addition to these lights, it must further display a steady white light through angles of 110° ahead and to each side in the horizontal plane. If it is not under command or it has stopped its engines, it must display the white light dead ahead and at angles of 110° each side, and a white navigation light at dead astern and to angles of 70° to each side.

In addition to the white lights while under this condition, it must display two red lights showing in all directions, suspended below the control car. The first must be at least 8 m below the control car, and the second at least 4 m below that. The red and green navigation lights are only to be displayed if the vessel is making headway.

Collision Avoidance at Night

When flying at night, collision avoidance depends upon which lights of another aircraft are seen, in which direction, and what their disposition is. When the direction of another aircraft is discussed, it will be referred to as its bearing. It should be noted that the bearing referred to is in relation to the aircraft one is flying, not in relation to the compass rose.

It is fair to say that any light from another aircraft that remains constant in its bearing is on a collision course. The exception to this is a single white light, which is then dependent on its luminosity. That is to say that if a single white light is seen on a bearing of 000°/360°, your aircraft is following another aircraft. If the light were to get brighter and larger, your aircraft would be going faster and a collision risk would exist. You would therefore need to take corrective action to overtake the other aircraft. If the light became dimmer and smaller, the preceding aircraft is going faster, and no collision risk exists.

If the constant white light is seen anywhere from 290° through 000° to 070°

and is increasing in brightness, avoidance action is necessary. If it gets dimmer, however, no avoidance action is required.

If a constant white light is seen anywhere from 070° through 180° to 290°, then, regardless of intensity, it can be regarded that no collision risk exists.

Various other light configurations are illustrated in Fig. 4-12.

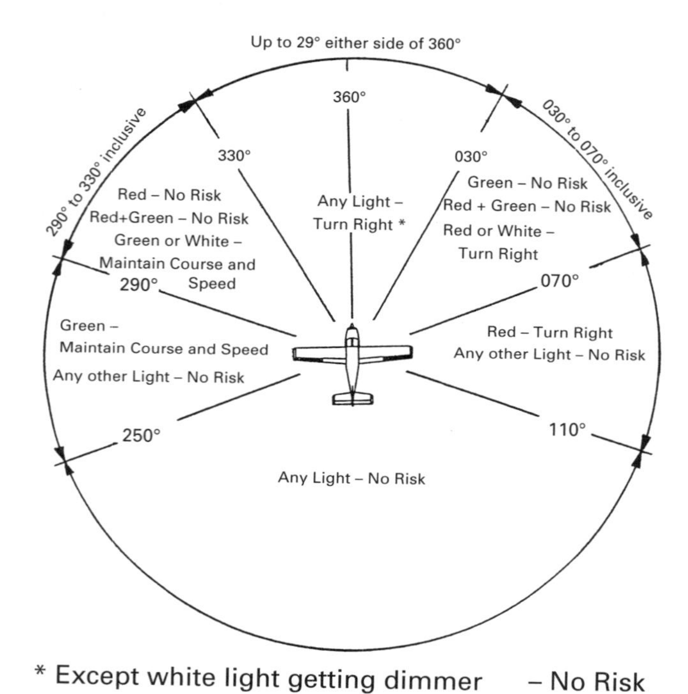

Fig 4–12 Collision Avoidance Rules

Low Flying

To permit safe flying and to avoid danger to persons and property on the ground, there are certain low-flying restrictions with relation to helicopters and aircraft other than helicopters.

With reference to helicopters, they must fly above a height that, in the event of an engine failure, will enable the machine to land safely without danger to persons or property. A helicopter is also restricted from flying over any congested area of a town, city or settlement below a height of 1,500 ft above the highest fixed obstacle within a 2,000 ft radius of the helicopter. Helicopters are also restricted from flying too low within a set area of London, this being specified and made clear in Rule 5 of the ANO.

When flying over congested areas, the lowest height at which other aircraft should fly is the greater of the following heights, which is either:

a 1,500 ft above the highest fixed obstacle within 2,000 ft, or
b A height which would enable the aircraft to fly clear of the area in the event of an engine failure. If the aircraft is towing a banner, this height will be calculated on the assumption that the banner will fall clear of the congested area.

An aircraft should not fly closer than 500 ft to any person, vessel or structure except under the following conditions:

a Take-off and landing
b A glider aircraft that is hill soaring
c A police aircraft
d Aircraft taking part in an air display
e Picking up and dropping of banners, tow ropes, etc.
f An organised contest or race.

Some of these exemptions require the written permission of the organisers of such events as the event requires.

Gatherings and Assemblies

Where there are organised ground events, whereby more than 1,000 persons are gathered together, an aircraft should not fly:

a over, or within 3,000 ft of the gathering unless written permission is given by the Authority,

OR

b below a height which would not enable the aircraft to fly clear of the area in the event of an engine failure. If the aircraft is towing a banner, this height will be based on the banner not being able to drop within 3,000 ft of the assembly.

If a flight was made, by accident, over or within 3,000 ft of an assembly, thereby contravening either of the above conditions, it is a justifiable defence if the pilot can prove that they are not connected with the gathering or assembly in any way, and that the flight was made at a reasonable height.

There are some further general exceptions to all the rules stated above. These are:

1 Aircraft that are involved in landing, taking off and practising approaches at aerodromes suitably designated
2 Not applicable to kites or captive balloons
3 Aircraft flying in a manner to save life
4 Not applicable to aircraft which have an Aerial Certificate granting permission for the picking up or dropping of articles.

Chapter 5

Aerodromes and Procedures

The information with reference to aerodromes is to be found in the *UK Air Pilot* AGA section, which lists all the aerodromes where landing by civil aircraft is permitted in the United Kingdom. Exceptions to this are where special permission has been granted by an aerodrome operator, or where a genuine emergency exists.

The AGA section listing the permitted aerodromes also contains the relevant information on each aerodrome, giving the characteristics of that particular aerodrome plus local warnings and obstacles. The various categories of aerodromes are divided into:

Government Aerodromes These include military aerodromes, aerodromes owned by other Government Departments and also those owned by the CAA. The CAA aerodromes are those licensed for 'public use', and are open for the period of time notified in the *UK Air Pilot*. Landing and take-off outside the published hours may only be possible with prior permission.

Aerodrome Licence (Ordinary) These are aerodromes licensed for use by the licensee or those who have prior permission. Therefore flights in to and out of these aerodromes require the prior permission of the owner or his delegate.

Unlicensed Aerodromes These aerodromes do not include abandoned aerodromes, but are unlicensed for public use. They are listed in the *UK Air Pilot*, but as it is the owners' responsibility to keep the CAA informed on facilities and current state, the information in the *UK Air Pilot* is not verified and may be out of date. Prior permission of the owner or his delegate is required before landing or taking off at these aerodromes.

Military Aerodromes

These are covered under the Government Aerodrome heading, but more information is required to be known. A list of military aerodromes which are available for use by civil aircraft are given in the *UK Air Pilot*. The information on these aerodromes is given to the CAA by the aerodrome, but the aerodrome is not inspected to obtain this information. These aerodromes and those listed in RAC 3-2 require prior permission for take-off and landing before the aircraft takes off from its original departure point, and are restricted to normal times of opening.

Foreign aircraft are not permitted to use these military aerodromes. However, certain aerodromes are used as diversion aerodromes in the event of adverse conditions preventing a landing at the original destination. These aerodromes are designated **MILITARY EMERGENCY DIVERSION AERODROMES (MEDA)**. When using military aerodromes, certain instructions are required to be complied with, and the facilities available and their use are provided under strict conditions.

To begin with, prior permission to land must be given before initial take-off, and for operational or administrative reasons the request may be denied. If this is the case, the rejection must be accepted as final, regardless of the reasons given. Once the aircraft has landed, and also before subsequent take-off, the commander must report to the ATC and give details of the aircraft, crew and passengers. All instructions regarding taxying, take-off and landing must be complied with, and any special procedures required are published in the aerodrome directory (AGA 3).

Apart from exceptional circumstances and cases of distress, the loading/unloading and servicing of civil aircraft will not be undertaken by military units. Fuel, oil and similar products may be supplied, but it is entirely at the discretion of the Station Commander.

Use of the aerodrome and any equipment will be entirely at the owner's risk, with no liability by that aerodrome. This is also the case where hangarage is supplied. Hangarage will only be supplied once all the military requirements are met.

Civilian pilots will be given the obstacle clearance height (OCL) based on the aerodrome operating minima contained in RAC 4-3. The QNH for that aerodrome will be passed to civilian pilots at all military aerodromes except certain Royal Naval Air Stations, listed in RAC 3-8-2-1, which still operate on QFE. A requirement for two-way RTF communication is essential, operating on 122.10 mHz, though very light aircraft without RTF may be treated exceptionally at the discretion of the Station Commander. Finally, at military aerodromes, civilian use is subject to a charge being made at the time.

The hours of operation of an aerodrome service will differ between aerodromes and between licensed and unlicensed aerodromes. The type of service given will depend on the type of aerodrome and the movements with which it has to cope. The type of service is annotated in the list in the *UK Air Pilot*, and is categorised as follows:

ATC Air Traffic Control Officers hold licences with appropriate ratings and are subject to inspection by the CAA.

AFIS Aerodrome Flight Information Service Officers hold licences and may be subject to inspection by the CAA.

A/G This service is not subject to CAA inspection, and therefore personnel providing these services are not required to hold ATC or AFISO licences.

NIL No service is provided.

Runway and Aerodrome Specifications

The elevation of the aerodrome as written in the AGA section of the *UK Air Pilot* is defined as being the highest point on the landing area, and is given as feet above mean sea level (AMSL). The exception to this is in Northern Ireland, where it is given in feet above the Belfast Bay Datum.

Runway Classifications

Airport runways are constructed of various materials, and these are indicated by codes, as follows:

A Asphalt
B Bitumen
C Concrete
D Gravel
G Grass
S Sand.

Each type of runway has a local bearing capacity. It is important to know whether or not your aircraft can safely land on an airport's runway without damaging the runway or the aircraft. The information on runway construction is provided by the Pavement Classification Number system found in the En route Supplement.

Runway Classification Number

The runway classification number system is a five-part code that provides the strength of a runway. The five-part code has numbers and letters (80/R/B/W/T) which provide:

a The pavement classification number (PCN)
b The type of pavement
c The pavement sub-grade category
d The maximum tyre pressure authorised for the pavement
e The pavement evaluation method.

PAVEMENT CLASSIFICATION NUMBER The PCN is the first number or numbers of the code. The PCN shows the load-bearing strength of a runway, and is calculated and assessed by a technical evaluation. The PCN value for each aircraft is listed in the Flight Information Handbook. If the PCN value for your aircraft is equal to or less than the runway PCN, it is safe for you to land on that runway.

TYPE OF PAVEMENT The type of pavement is expressed in either of two terms, which are:

R Rigid
F Flexible.

PAVEMENT SUB-GRADE CATEGORY The pavement sub-grade relates to the material that is placed under the runway surface material to act as a base. The pavement sub-grade material is categorised according to its ability to withstand landing loads. The categories and code for each are as follows:

Category	**Code**
High	A
Medium	B
Low	C
Ultra-low	D

MAXIMUM TYRE PRESSURE AUTHORISED Tyres with high pressure exert more pressure per square inch on a runway surface than those with low pressure. Some runways therefore have tyre pressure limitations to prevent their being damaged by high-pressure tyres.

The maximum tyre pressure authorised is the fourth part of the PCN, and is shown as follows:

Maximum Tyre Pressure	**Code**
High, no limit	W
Medium, 217 psi limit	X
Low, 145 psi limit	Y
Very low, 73 psi limit	Z

PAVEMENT EVALUATION METHOD The fifth and final part of the five-part PCN code is the method used to evaluate the pavement. There are two evaluation methods, which are:

a Technical evaluation using engineering data and equipment. The code for this type of evaluation is 'T'
b Evaluation based on the experience of aircraft that have used the actual pavement. The code for this type of evaluation is 'U'.

EXAMPLE OF A PCN An example of a PCN as would be seen in the En route Supplement is 70/R/B/Y/T. These PCN codes show that the runway:

a Has a PCN of 70, and only aircraft with a PCN value of 70 or less can land on it
b Has a rigid (R) construction
c Has a runway sub-grade material (B) of medium strength
d Has a maximum tyre pressure limit of 145 psi or less
e Was evaluated technically (T).

The PCN is normally used in conjunction with the Aircraft Classification Number (ACN), and together these form a pavement bearing strength classification system for aircraft above 5,700 kg maximum take-off weight authorised (MTWA). The actual ACN is calculated by taking into account:

a aircraft weight
b pavement type
c sub-grade category.

For runways intended for use by aircraft weighing 5,700 kg or less, the bearing strengths are given as the maximum allowable tyre pressure. The operating procedures for the ACN/PCN system is to be found in AGA 7-2.

Aerodrome Distance Declarations

For an aerodrome licence to be issued, many requirements have to be met by the aerodrome operator. Many runways have usable and unusable sections, and various other criteria based on ACN/PCN requirements. Runway usable distances and take-off distances must be declared to the CAA for the issue of a licence. These 'DECLARED DISTANCES' are as follows and illustrated in Fig. 5-1.

TAKE-OFF RUN AVAILABLE (TORA) This is defined as the length of the runway which is available and suitable for the ground run of an aeroplane taking off.

EMERGENCY DISTANCE AVAILABLE (EDA) This is defined as the length of the declared TORA plus the length of the stopway available.

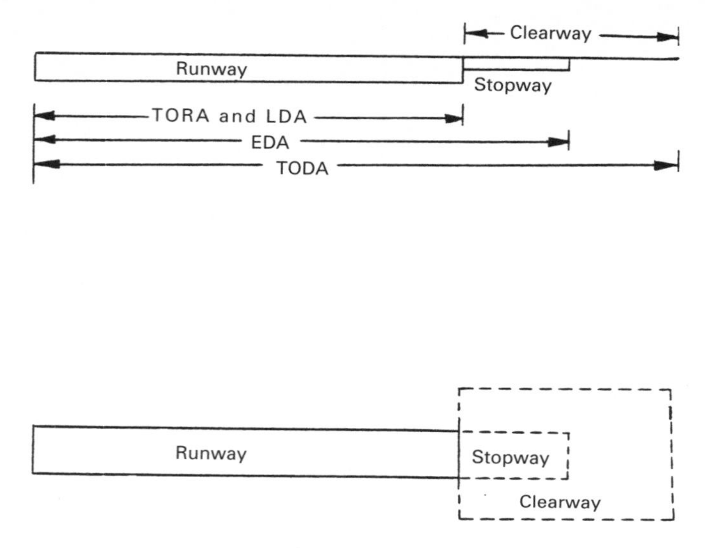

Fig 5–1 Aerodrome Declared Distances

TAKE-OFF DISTANCE AVAILABLE (TODA) This is defined as the length of the declared take-off run plus the length of clearway available, and will not exceed 1.5 TORA.

LANDING DISTANCE AVAILABLE (LDA) This is defined as the length of runway (or surface, when this is unpaved) available and suitable for the ground landing run of the aeroplane, commencing at the landing threshold or displaced landing threshold.

STOPWAY An area ahead of the runway, clear of upstanding obstacles and capable of supporting the weight of the aircraft. It must have a braking coefficient as good as, if not better than, the TORA.

CLEARWAY A defined rectangular area at the end of a strip or channel in the direction of selected take-off, or prepared as a suitable area over which an aircraft may make its initial climb to a specified height (ICAO). This information is therefore only for the use of departing aircraft.

Paved Runway Markings

Paved runways may have a variation of markings, dependent upon the type of runway (as illustrated in Fig. 5-2). However, there is a minimum scale of

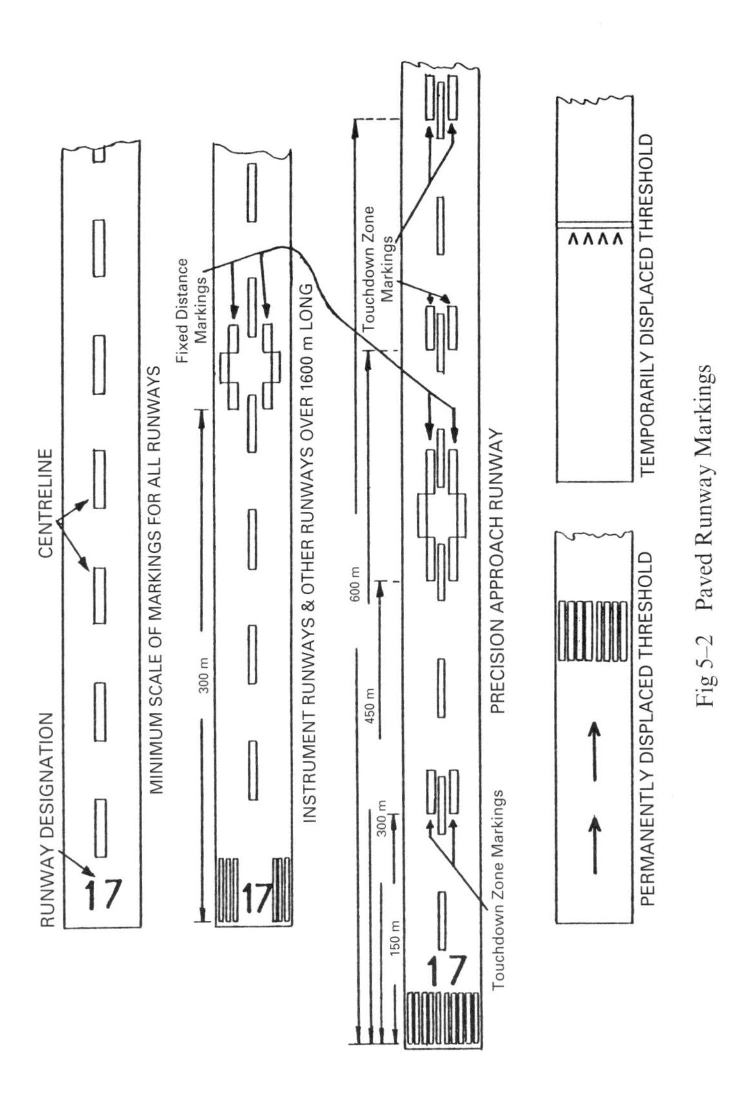

Fig 5–2 Paved Runway Markings

runway markings, which consists of the runway designation markings and runway centreline markings. The runway threshold may be marked by the designation number alone, or may have parallel stripes in line with the runway direction.

The landing threshold may or may not be at the end of the runway, depending on circumstances. When the threshold is not at the end of the runway it is known as a 'DISPLACED THRESHOLD', and may or may not be permanent. A displaced threshold may be temporary when repair work is being completed at the end of the runway. There could be various other reasons for a temporary displaced threshold, as illustrated in Fig. 5-2. A permanent displaced threshold may be seen and used where the ACN at the end of the runway is not strong enough for landing but is strong enough to support the weight of an aircraft. A permanently displaced threshold marking is illustrated in Fig. 5-2.

All precision approach runways which are 1,800 m and over and do not have a visual approach slope guidance system have additional markings called 'FIXED DISTANCE MARKINGS' set at 300 m from the threshold. The area between the threshold markings and the fixed distance marking is known as the 'OPTIMUM TOUCHDOWN ZONE'. Further markings may be found on runways with precision approach aids, and may extend for distances up to 900 m from the threshold. These are known as 'TOUCHDOWN ZONE MARKINGS' and are illustrated in Fig. 5-2.

Paved Taxiway Markings

Like the runways, the taxiways also have markings to indicate direction and centreline, as well as holding points. These markings are yellow, although some airfields may retain the old-type white markings. The taxi holding position markings on all UK airfields are to be updated to the new standard for ICAO. The new taxi holding markings are of two types, depending on the type of approach category. In all cases, a double solid and double pecked yellow lines will be indicated, and will represent visual/Cat 1 and Cat 2/Cat 3 markings when only one taxi holding position is available.

Where two taxi holding position markings are available, the double solid/double pecked are for the visual/CAT 1 hold, and are the closest to the runway. In this instance a CAT 2/Cat 3 hold markings are farthest from the runway and are indicated by double solid yellow lines with 'cross-bars'. Where two taxiways cross each other, taxiway intersection markings are shown. These consist of a single pecked yellow line across the taxiway. These markings are illustrated for clarification in Fig. 5-3.

To enable aircraft to manoeuvre on to the apron and in to set stand markings, yellow solid lines guide the aircraft into their particular stands, which are numbered accordingly. There are several variations of these within the UK, but they are clearly marked and, when using them, it is important to remain on the centreline.

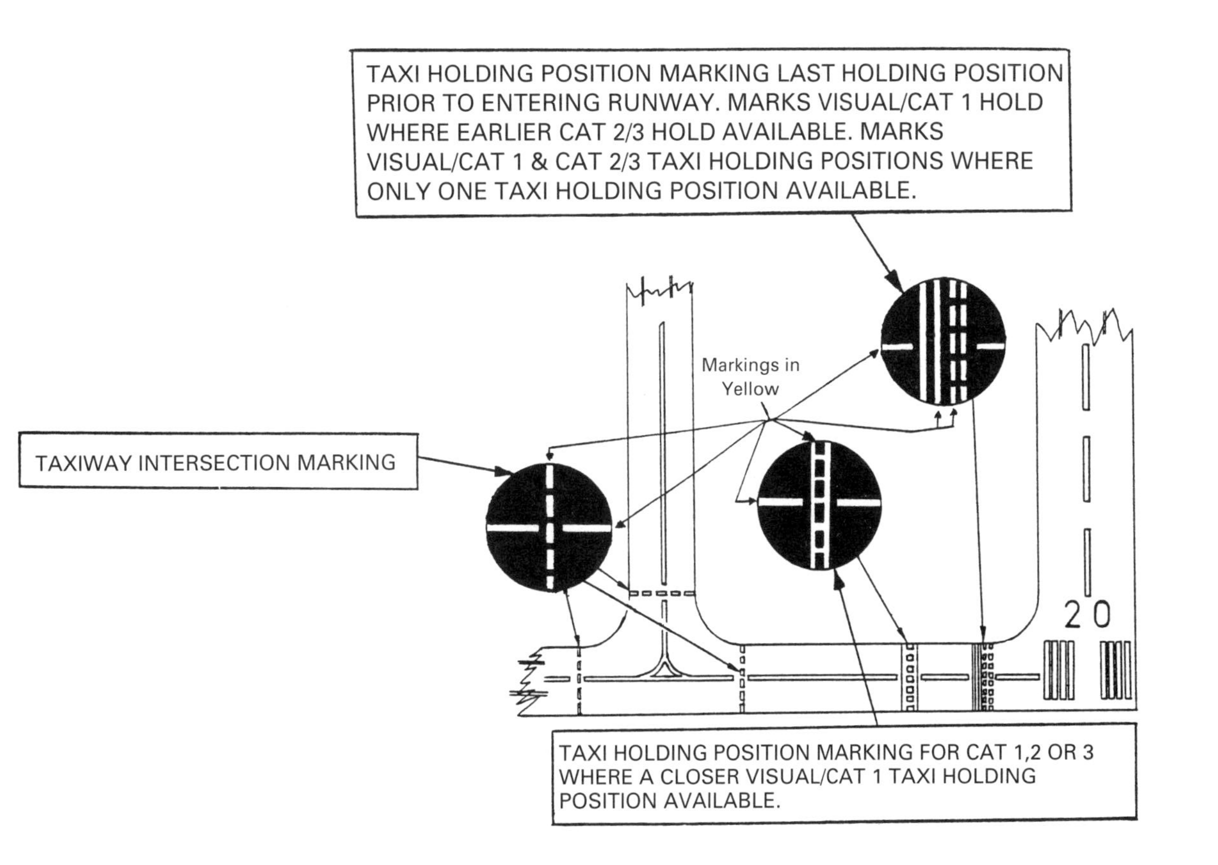

Fig 5–3 Paved Taxiway Markings

Aerodrome Signals and Markings

a A white landing 'T' signifies that aeroplanes and gliders taking off or landing shall do so in a direction parallel with the shaft of the 'T' and towards the cross arm, unless otherwise authorised by the appropriate air traffic control unit

b A white disc 60 cm in diameter displayed alongside the cross arm of the 'T' and in line with shaft of the 'T' signifies that the direction of landing and take-off do not necessarily coincide

c A white dumbbell signifies that movement of aeroplanes and gliders on the ground shall be confined to paved, metalled or similar surfaces

d A white dumbbell with a black strip 60 cm wide across each disc at right angles to the shaft of the dumbbell signifies that the aeroplanes and gliders taking off or landing do so on a runway but that movement on the ground is not confined to paved, metalled or similar hard surfaces

e A red and yellow striped arrow as shown signifies that a right-hand circuit is in force

f A red square, 3 m square, with a yellow strip diagonally across, signifies that the state of the manoeuvring area is poor and pilots must exercise special care when landing

g A red square, 3 m square, with a yellow cross from corner to corner as indicated, signifies that the aerodrome is unsafe for the movement of aircraft and that landing is prohibited

h A white letter 'H' signifies that helicopters shall take off and land only within the area designated by the 'H' marking

i A white double cross signifies that glider flying is in progress

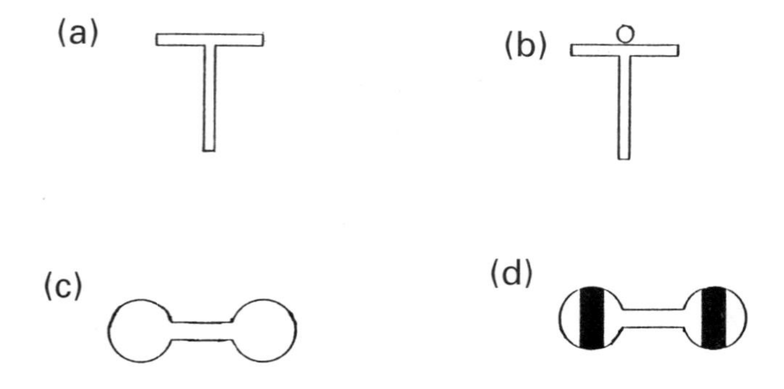

Fig 5–3a Aerodrome Signals and Markings

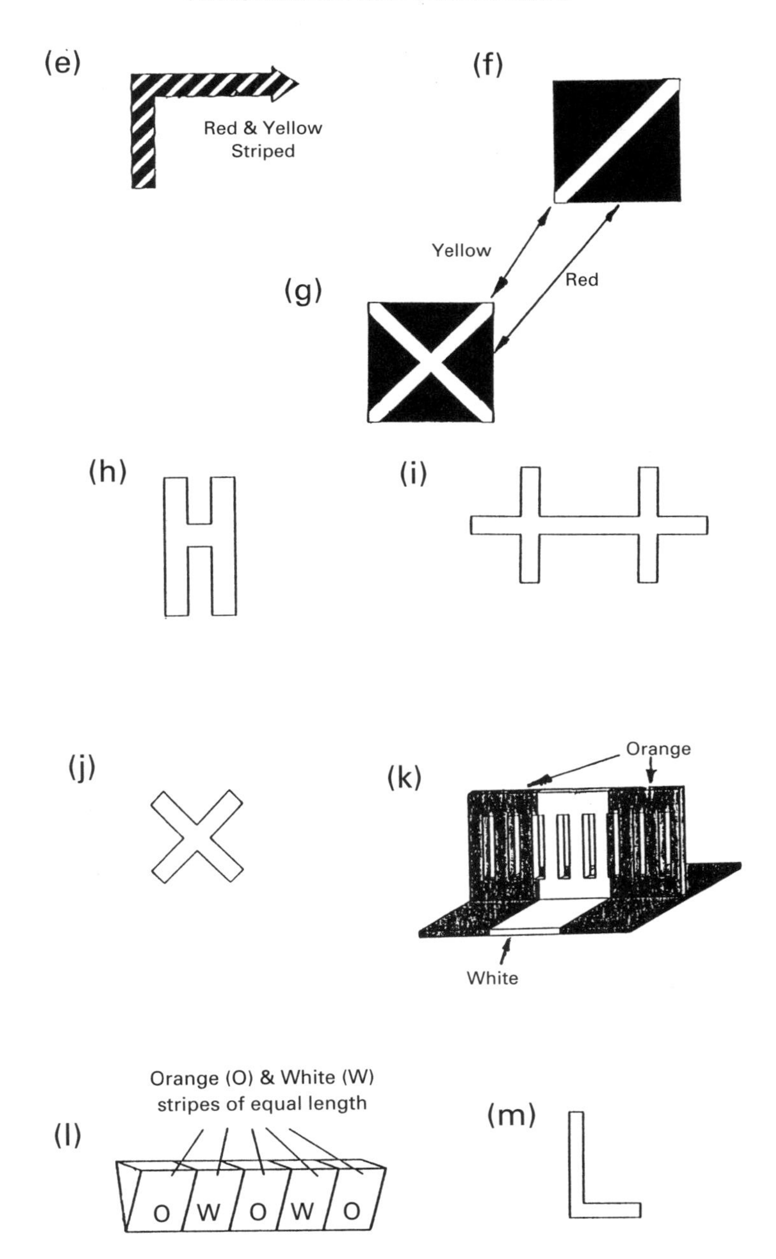

Fig 5–3b Aerodrome Signals and Markings (Cont'd)

j Two or more white crosses displayed on a runway or taxiway signify that the section of the runway or taxiway marked by them is unfit for the movement of aircraft

k Orange and white markers spaced not more than 15 m apart signify the boundary of that part of the paved runway, taxiway or apron which is unfit for the movement of aircraft

l Markers with orange and white stripes as indicated indicate the boundary of an area unfit for the movement of aircraft, the said area being indicated by one or more white crosses

m A white letter 'L' indicates a part of the manoeuvring area which shall be used for the taking off and landing of light aircraft only.

Marshalling Signals

(See Fig 5–3c)

Aerodrome Procedures

Altimeter Setting Procedures

The altimeter is a pressure differential sensing instrument which senses the difference between the pressure setting in the instrument and the pressure at the flight altitude. The difference between the two pressures is mechanically converted into an altitude indication. To provide an accurate altitude indication, the current barometric pressure must be set into the altimeter and the accuracy of the altimeter checked. This will provide suitable pressure information which will assist the pilot in maintaining adequate terrain clearance and ensure a safe standard of flight separation.

The altimeter is designed to indicate the correct altitude for standard atmosphere temperature and pressure. Standard temperature is a temperature at sea level of 15°C and a decrease in temperature of 2°C for each 1,000 ft of altitude. The standard atmospheric pressure at sea level is 29.92 in of mercury (in HG), which is 1013.2 mb, (see conversion tables in Table 5-1, page 71), and the pressure decreases as altitude increases. When temperatures and pressures differ from the standard, the indicated altitude is not correct. A change in the atmospheric pressure of 1 in of mercury can result in a change of approximately 1,000 ft in the altitude indication. A change of 1 mb is equal to approximately 30 ft in altitude indication. The altimeter must therefore be adjusted for the current barometric pressure so that the altitude indications are correct.

TRANSITION Transition is the general term that describes the change from one phase of flight or flight condition to another (from en route flight to the approach, or from instrument flight to visual flight, for example).

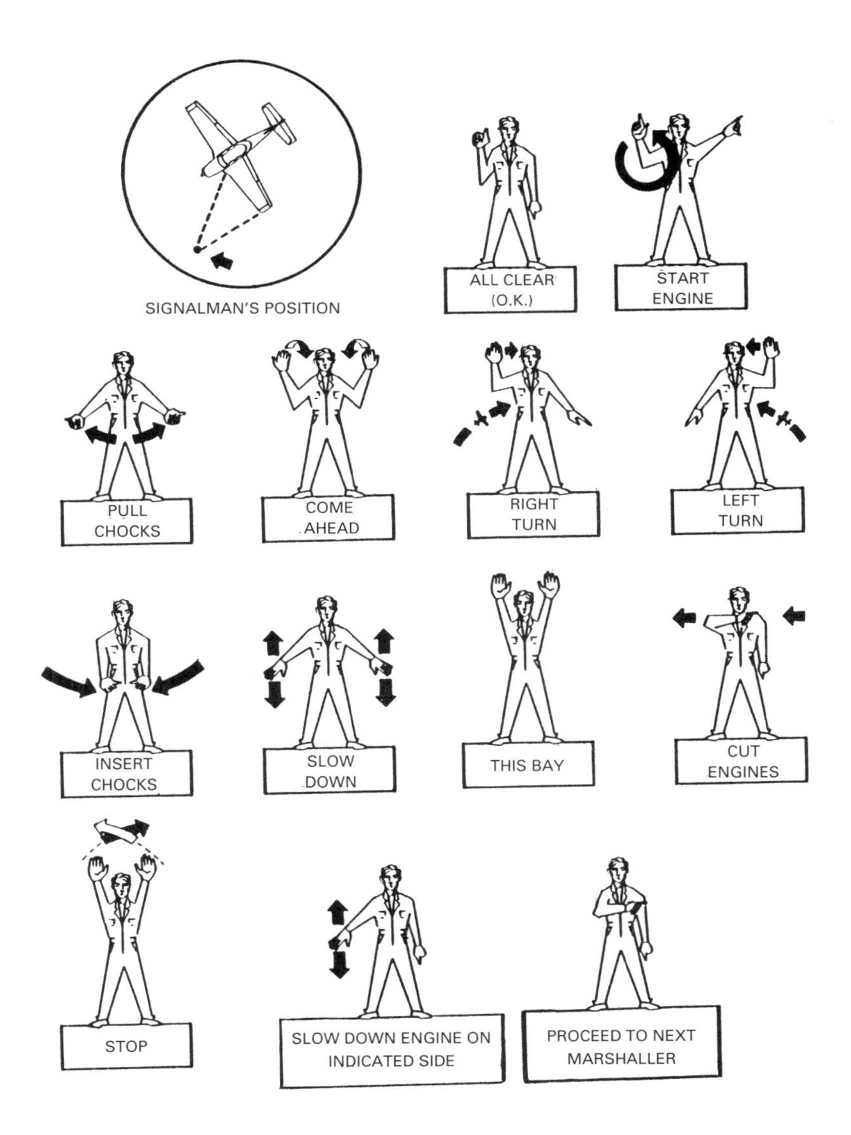

Fig 5–3c Marshalling Signals

TERMINOLOGY The following terminology is associated with altimeter pressure settings and flight conditions, and is shown in Fig. 5-4:

1 ALTITUDE The vertical distance of a level, point or an object, considered as a point measured from mean sea level
2 ELEVATION The vertical distance of a level or point fixed to the surface of the Earth, measured from mean sea level
3 HEIGHT The vertical distance of a level or point, measured from the Earth's surface or specified datum
4 FLIGHT LEVEL Surfaces of constant atmospheric pressure which are related to a specific pressure datum of 1013.2 mb or 29.92 in HG
5 QNE The term related to the International Standard Atmospheric (ISA) pressure of 1013.2 mb or 29.92 in HG
6 QNH An airfield or local/regional pressure setting relating to altitude above mean sea level
7 QFE A specified airfield pressure setting relating to height above ground level of that airfield.

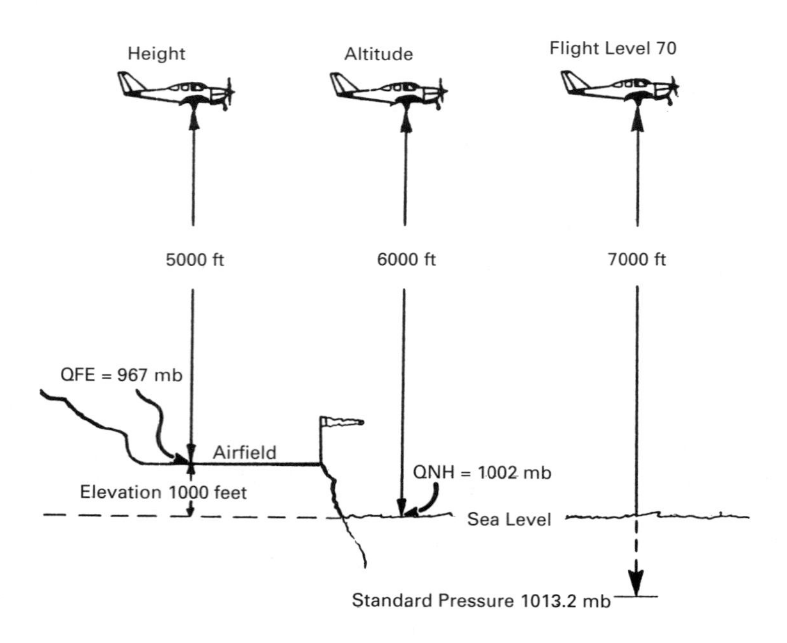

Fig 5–4 Terminology

Note: A pressure-type altimeter calibrated in accordance with the standard atmospheric pressure:

a When set to a QNH altimeter setting will indicate ALTITUDE
b When set to a QFE altimeter setting will indicate HEIGHT above the QFE reference airfield
c When set to a pressure of 1013.2 mb or 29.92 in HG will indicate a FLIGHT LEVEL.

Table 5-1

Conversion Tables

Barometric Pressure Readings

Inches	hPa/mbs	Inches	hPa/mbs	Inches	hPa/Mbs	Inches	hPa/mbs	Inches	hPa/mbs
28.00	948.2	28.29	958.0	28.58	967.8	28.87	977.7	29.16	987.5
01	948.5	28.30	958.3	59	968.2	88	978.0	17	987.8
02	948.9	31	958.7	28.60	968.5	89	978.3	18	988.2
03	949.2	32	959.0	61	968.8	28.90	978.7	19	988.5
04	949.5	33	959.4	62	969.2	91	979.0	29.20	988.8
05	949.9	34	959.7	63	969.5	92	979.3	21	989.2
06	950.2	35	960.0	64	969.9	93	979.7	22	989.5
07	950.6	36	960.4	65	970.2	94	980.0	23	989.8
08	950.9	37	960.7	66	970.5	95	980.4	24	990.2
09	951.2	38	961.1	67	970.9	96	980.7	25	990.5
28.10	951.6	39	961.4	68	971.2	97	981.0	26	990.9
11	951.9	28.40	961.7	69	971.6	98	981.4	27	991.2
12	952.3	41	962.1	28.70	971.9	99	981.7	28	991.5
13	952.6	42	962.4	71	972.2	29.00	982.1	29	991.8
14	952.9	43	962.8	72	972.6	01	982.4	29.30	992.2
15	953.3	44	963.1	73	972.9	02	982.7	31	992.6
16	953.6	45	963.4	74	973.2	03	983.1	32	992.9
17	953.9	46	963.8	75	973.6	04	983.4	33	993.2
18	954.3	47	964.1	76	973.9	05	983.7	34	993.6
19	954.6	48	964.4	77	974.3	06	984.1	35	993.9
28.20	955.0	49	964.8	78	974.6	07	984.4	36	994.2
21	955.3	28.50	965.1	79	974.9	08	984.8	37	994.6
22	955.6	51	965.5	28.80	975.3	09	985.1	38	994.9
23	956.0	52	965.8	81	975.6	29.10	985.4	39	995.3
24	956.3	53	966.1	82	976.0	11	985.8	29.40	995.6
25	956.7	54	966.5	83	976.3	12	986.1	41	995.9
26	957.0	55	966.8	84	976.6	13	986.5	42	996.3
27	957.3	56	967.2	85	977.0	14	986.8	43	996.6
28	957.7	57	967.5	86	977.3	15	987.1	44	997.0

Conversion Tables(Cont'd)

Inches	hPa/mbs	Inches	hPa/mbs	Inches	hPa/Mbs	Inches	hPa/mbs	Inches	hPa/mbs
45	997.3	29.76	1007.8	30.07	1018.3	30.38	1028.8	30.69	1039.3
46	997.6	77	1008.1	08	1018.6	39	1029.1	30.70	1039.6
47	998.0	78	1008.5	09	1019.0	30.40	1029.5	71	1040.0
48	998.3	79	1008.8	30.10	1019.3	41	1029.8	72	1040.3
49	998.6	29.80	1009.1	11	1019.6	42	1030.1	73	1040.6
29.50	999.0	81	1009.5	12	1020.0	43	1030.5	74	1041.0
51	999.3	82	1009.8	13	1020.3	44	1030.8	75	1041.3
52	999.7	83	1010.2	14	1020.7	45	1031.2	76	1041.7
53	1000.0	84	1010.5	15	1021.0	46	1031.5	77	1042.0
54	1000.4	85	1010.8	16	1021.3	47	1031.8	78	1042.3
55	1000.7	86	1011.2	17	1021.7	48	1032.2	79	1042.7
56	1001.0	87	1011.5	18	1022.0	49	1032.5	30.80	1043.0
57	1001.4	88	1011.9	19	1022.4	30.50	1032.9	81	1043.3
58	1001.7	89	1012.2	30.20	1022.7	51	1033.2	82	1043.7
59	1002.0	29.90	1012.5	21	1023.0	52	1033.5	83	1044.0
29.60	1002.4	91	1012.9	22	1023.4	53	1033.9	84	1044.4
61	1002.7	92	1013.2	23	1023.7	54	1034.2	85	1044.7
62	1003.1	93	1013.5	24	1024.0	55	1034.5	86	1045.0
63	1003.4	94	1013.9	25	1024.4	56	1034.9	87	1045.4
64	1003.7	95	1014.2	26	1024.7	57	1035.2	88	1045.7
65	1004.1	96	1014.6	27	1025.1	58	1035.5	89	1046.1
66	1004.4	97	1014.9	28	1025.4	59	1035.9	30.90	1046.4
67	1004.7	98	1015.2	29	1025.7	30.60	1036.2	91	1046.7
68	1005.1	99	1015.6	30.30	1026.1	61	1036.6	92	1047.1
69	1005.4	30.00	1015.9	31	1026.4	62	1036.9	93	1047.4
29.70	1005.8	01	1016.3	32	1026.7	63	1037.3	94	1047.8
71	1006.1	02	1016.6	33	1027.1	64	1037.6	95	1048.1
72	1006.4	03	1016.9	34	1027.4	65	1037.9	96	1048.4
73	1006.8	04	1017.3	35	1027.8	66	1038.3	97	1048.8
74	1007.1	05	1017.6	36	1028.1	67	1038.6	98	1049.1
75	1007.5	06	1018.0	37	1028.4	68	1038.9	99	1049.5

TRANSITION ALTITUDE Transition altitude is the altitude in the vicinity of an airport at or below which the vertical position of an aircraft is controlled by reference to ALTITUDE. At or below the transition altitude, the altimeter pressure setting is set to the local QFE or QNH as given by ATC. When climbing out en route, the aircraft passes a level at which the altimeter setting is changed from the local QFE or QNH to the standard ISA setting of 1013.2 mb (or 29.92 in.HG – QNE). This level is the **transition altitude**, as indicated in Fig. 5-5. In the climb below and up to transition altitude the altitude is expressed in feet, and as a Flight Level above it.

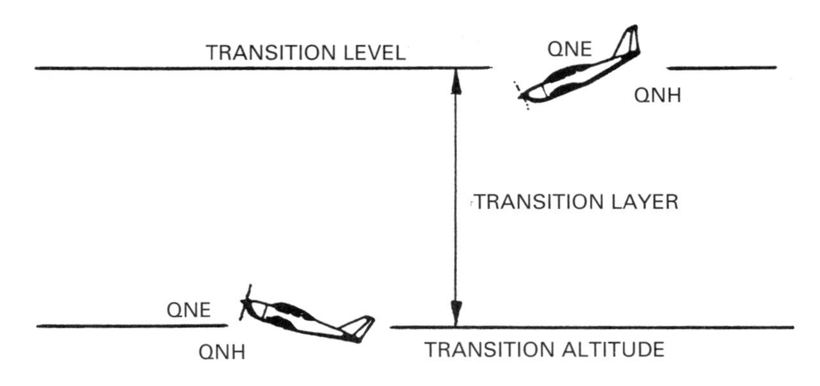

Fig 5–5 Transition Points

TRANSITION LEVEL AND LAYER The transition level is the lowest Flight Level available for use above the transition altitude. On descent into an aerodrome, a level is reached where the altimeter pressure setting is changed from the ISA standard to the aerodrome QFE or QNH. This level is the **transition level**, and may be regarded as the lowest Flight Level at which an aircraft can safely proceed on the en route altimeter setting. The airspace between transition altitude and transition level is known as the **transition layer**.

NOTE: *Transition altitude in the United Kingdom is 3,000 ft, except in or beneath certain airspace as nominated in RAC 2-2 of the* UK Air Pilot.

Aerodrome Circuit Traffic Patterns

Aerodrome traffic patterns ensure a safe and orderly flow of aircraft traffic. Take-off patterns are established to permit aircraft to take-off safely from the active runway and to either remain in or depart from the aerodrome traffic pattern. Landing traffic patterns are established to permit aircraft to enter the flow of the aerodrome traffic and continue to fly to a safe landing.

Traffic patterns can be either left-hand or right-hand. In a left-hand traffic pattern all turns are to be left-hand turns, and this is termed a STANDARD traffic pattern at most aerodromes. In a right-hand traffic pattern all turns are to the right; this is termed a NON-STANDARD traffic pattern and is normally used under special conditions (e.g. obstruction clearance).

CIRCUIT TRAFFIC PATTERNS An aerodrome circuit traffic pattern consists basically of four separate segments known as 'legs' (Fig. 5-6) which comprise:

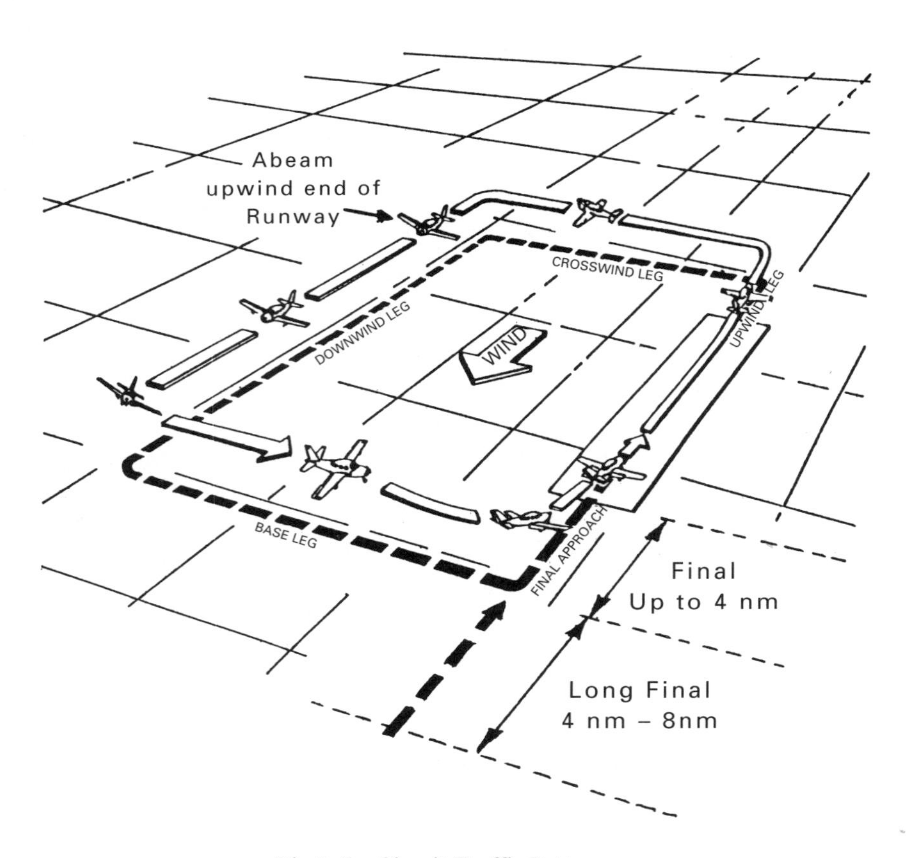

Fig 5–6 Circuit Traffic Pattern

a Crosswind leg – this is flown perpendicular to the take-off segment
b Downwind leg – this is flown parallel to the landing runway in a direction opposite to the landing, and normally extends between the crosswind leg and the base leg. Pilots are to report 'Downwind' when abeam the upwind end of the runway
c Base leg – this is used to transition the aircraft from the downwind leg to the final approach segment. The base leg is entered by making a 90° turn to the left (for a standard circuit) from the downwind leg to a flight path

that is perpendicular to the downwind leg. Aircraft are to report 'Base leg', if requested by ATC, immediately on completion of the turn on to a base leg

d Upwind/Final leg – the final leg is the final approach in line with the runway, and aircraft are to report 'Final' after the completion of the turn on to final approach and when at a range of not more than 4 nm from the approach end of the runway. Aircraft flying a final approach of a greater length than 4 nm are to report 'Long final' when beyond that range, and 'Final' when a range of 4 nm is reached. Aircraft flying a straight-in approach are to report 'Long final' at 8 nm from the approach end of the runway, and 'Final' when a range of 4 nm is reached. On take-off the aircraft is climbing in the 'Upwind' end of the runway leg, and can either continue the climb away from the circuit or can climb to the circuit height and turn 90° on to the 'Crosswind leg'.

CHAPTER 6

AERONAUTICAL INFORMATION SERVICES

Introduction

Various documents are available to provide aeronautical information for the pilot. However, certain documents are essential for the pilot's safety and the safety of other aviators. Although the ANO is one of these documents, it provides the legal aspect of flying with reference to Rules of the Air, the type of flying allowed and the articles that govern all forms of flying, from the balloonist and glider pilot up to flying a large transport aircraft. The ANO will be discussed separately in a later chapter.

The day-to-day information is supplied through the following documents:

a Aeronautical Information Publication (Air Pilot) CAP 32
b Aeronautical Information Circulars
c Notices to Airmen (NOTAMS)

These are discussed individually in the following text.

Aeronautical Information Publication (AIP) – CAP 32

More commonly known as the *UK Air Pilot*, this contains information of a practical operational nature, and its layout is of a standard applied throughout all the member countries of ICAO. This is an official publication which is divided into eight sections which relate to specific subjects, and therefore can be found in the same format in the AIP of any ICAO member country. These specific subjects range from aerodrome information and meteorology to search and rescue, as well as other related subjects in the flying spectrum, and refer to other legislation that is not necessarily laid down in the ANO.

The eight sections in the AIP are described as follows:

GEN This general section contains matters of an administrative nature, including a list of abbreviations, a conversion table, a note on the time system, public holidays, a guide to the NOTAM distribution system and a list of Civil Aviation Legislation.

AGA The AGA section contains all the information relating to aerodromes, inclusive of Government aerodromes, licensed aerodromes (both public and ordinary) and unlicensed aerodromes. This information includes detailed maps of the aerodrome layout, the classifications and limitations on their use, the types of available ground services, times of opening and operation, runway directions, airfield elevation, transition altitude, visual ground aids, aerodrome obstacles and obstructions, movement areas and declared distances.

COM The abbreviation COM is short for communications, and therefore the COM section deals with this subject. Four basic services are covered in this section, namely:

a Aeronautical Fixed Service
b Aeronautical Mobile Service
c Aeronautical Radio Navigation Service
d Aeronautical Broadcast Service.

These four services are described and outlined in this section, followed by an alphabetical list of navigational radio aids identification. A listing of various radio facilities at various aerodromes is covered under the heading 'Radio Communication and Navigation Facilities'. En route and long-range facilities for navigation and communications, along with the appropriate frequencies used on airways, are also included, as well as the broadcasting of radio time signals.

MET The MET section outlines the organisation and the distribution network for the meteorological service. The detail for information and operating procedures, with reference to weather conditions for aircraft on air routes, can also be obtained from this section. The various tabled appendices describe how various met forecasts can be decoded, and the definitions associated with these forecasts. The areas of responsibility and the centres associated with each watch area are outlined, and the addresses and telephone numbers for use by pilots requiring information are provided.

RAC This section covers the operational and safety aspect of flying, including the rules of flight, the types of airspace and the control of flying within them, the filing of flight plans, the requirements covering carriage of radio, airmiss reporting, danger and prohibited areas, altimeter setting requirements and setting regions. The differences between international and UK procedures are also listed, as well as definitions.

FAL The administration and requirements of procedures covering arrivals, departures and transit within the UK and overseas destinations. This covers Customs, health, documentation of passengers, and the export and import of cargo.

SAR The search and rescue organisation, along with emergency procedures and international signalling codes, the facilities available to pilots, and the alerting, communications and general procedures are all covered in this section.

MAP This section describes the aeronautical charts and maps available, the aeronautical symbols depicted and their meaning, and provides a list of charts and any corrections to be made to them.

A monthly amendment and supplement service keeps the *UK Air Pilot* up to date with any changes that occur. It is up to the individual to amend these changes by removing the corrected pages, destroying them, and replacing them with the updated amendment.

Amendments to the AIP

Amendments to the AIP take the form of AIRAC or NON-AIRAC amendments. AIRAC amendments cover changes of a significantly operational nature, such as airspace changes and description. NON-AIRAC amendments cover non-operationally significant changes. All amendments are issued monthly, and in both cases the cover sheet issued with the amendments will briefly describe all NOTAMS that have been incorporated.

Aeronautical Information Circulars (AICs)

The AICs are leaflets which are not generally of an operational nature, and are issued monthly on subjects which concern the administrative aspect of flying. Impending operational changes, and advance warnings, are sometimes published in an AIC to enable aircraft operators to make necessary arrangements. Through the medium of these circulars, information regarding matters of an operational nature, and corrections to charts and amendments can be made available to all operators and flight crew personnel. Different-coloured paper is used to indicate the subject matter, the classifications being:

White Administrative matters; e.g. examination dates, new publications or amendments to publications, courses fee/charges.

Yellow	Operational matters including ATS facilities and requirements.
Pink	Matters which need special emphasis on safety.
Mauve	Amendments to UK airspace restrictions charts.
Green	Maps and charts.

They are valid for five years, after which period they will lapse automatically on their fifth anniversary, unless amended before that time. If the subject matter is found to be of continual momentous importance, they will be scrapped and reissued under a new date and AIC reference number.

For holders of the AIS integrated package, the monthly amendments are part of the service, and are issued free of charge. For non-AIP holders, the current cost for postage and packing is £10 a year from January until December.

NOTAMS

These are notices which cover all operational information which is NOT covered by the AIP Amendment or Supplement service. The name is derived from the term Notices to Airmen. NOTAMS are issued by the Aeronautical Fixed Telecommunication Network (AFTN), and cover changes, both permanent and temporary, which are of operational significance and need to be introduced at short notice. The AIP amendments and supplements will supersede the NOTAMS at the earliest opportunity and as required.

NOTAMS with operationally significant changes are known as TRIGGER NOTAMS, and these will subsequently be issued as AIP Supplements or Amendments. The TRIGGER NOTAMS act as reminders which should be included in the appropriate pre-flight information bulletins.

The effective date of a TRIGGER NOTAM is on the date of the change, and it will remain valid for the duration of temporary changes, and for 15 days in the case of permanent changes.

The Aeronautical Information Service

The organisation of the Aeronautical Information Service (AIS) is part of the International Service, and forms part of the National Air Traffic Service (NATS) within the UK. The service within the AIS operations section is a 24-hour service, and is based at its central office at Heathrow Airport. It is responsible for initiating NOTAMS to all subscribers.

CHAPTER 7

USE OF RADAR IN AIR TRAFFIC SERVICES

Introduction

The use of radar in ATC is paramount in aviation, inasmuch as it increases the safety of flying and allows separation standards to be maintained, enabling collision avoidance to be effected. This radar surveillance can cover the whole of the UK airspace, and is known as AREA RADAR, or can be specific to individual aerodromes or groups of aerodromes, when it is known as APPROACH RADAR.

Area radar is a joint radar unit comprising both civil and military units, and is divided into levels of airspace comprising surveillance radar which can be with or without height-finding equipment.

The approach radar is individual to an aerodrome or airport with surveillance radar and varying degrees of precision approach radar.

The density of air traffic invites the use of SSR, which is associated with both of the above-mentioned units and is normally used to identify the air traffic concerned.

The procedures for the use of radar in ATS are laid down by the ICAO, and the UK generally subscribes to these. Although radar is of paramount importance in the control of air traffic in controlled airspace, it is also widely used to maintain separation of air traffic in uncontrolled airspace. The Radar Control Service is operated in airspace Classes A to E, while Class F and G airspace is provided with either a Radar Advisory Service (RAS) or a Radar Information Service (RIS). In simple terms, the Radar Control Service is used in controlled airspace and the Radar Advisory and Radar Information services are used in uncontrolled airspace. These classes of airspace and types of service are set out in the following table.

Table 7.1 Operation of Radar Services

Type of Airspace	Type of Service	ATC action with regard to Unknown Aircraft
Class A airspace, controlled airspace subject to IFR at all times and Class D airspace, controlled airspace below FL245 in which all flights are subject to the ATC authority	Radar Control Service	Traffic information and avoiding action will not be given unless information has been received which indicates that a radar echo may be an aircraft which is lost or experiencing radio failure
Class E airspace, controlled airspace in which with VFR, flight without ATC clearance is permitted.	Radar Control Service	Traffic information will be passed provided this does not compromise radar sequencing of traffic or separation of IFR flights. Avoiding action will be given at the request of pilots, but to limits decided by the radar controller, or if information has been received which indicates that a radar echo may be a particular aircraft which is lost or experiencing radio failure
Class B airspace, Upper Airspace Control Area	Radar Control Service	**(1) Within MRSAs:** (*p.89*) Generally, all aircraft operating off the promulgated Upper ATS routes will be vectored clear of those operating on the routes. To eliminate the possibility of a radar induced conflict, neither traffic information nor avoiding action will be given unless information received indicates an unknown aircraft is lost or has radio failure

Table 7.1 (Cont'd) **Type of Airspace**	**Type of Service**	**ATC action with regard to Unknown Aircraft**
		(2) Outside MRSAs: Whenever practicable, traffic information will be given. Avoiding action will be given if ATC consider it is necessary or if it is requested by pilots **Note:** Owing to the sudden appearance and the unpredictable movement of unknown aircraft, the required separation provisions are not always possible
Class F airspace, Advisory Routes	Radar Advisory Service or Radar Information Service	Traffic information will be passed, followed by advice on avoiding action
Class G airspace, all other airspace		Traffic information will be passed, but no avoiding action is to be given. Pilots are responsible for their own separation

The radar service is generally covered by the LARS or the MARS. Between them they cover all the airspace from ground level up to FL240. Class B airspace covers the Upper Airspace Control Area, and therefore from FL245 this airspace is covered by the MRSAs.

Outside Controlled Airspace

Radar Advisory Service

The RAS covers both Class F and G airspace, which is the advisory routes and all uncontrolled airspace. Although this service is generally available, it is dependent upon the workload of the controllers, and is therefore not an automatic right of the pilot. Traffic information for both Class F and Class G airspace is different, and is indicated in Table 7.1.

The RAS is an air traffic service whereby the controller will provide the necessary advice to maintain standard separation between participating aircraft, and in which he will pass to the pilot:

a The bearing
b The distance
c The level of conflicting non-participating traffic (if known),

together with advice on the necessary action to resolve the confliction. The controller will pass advice on avoiding action, followed by the information on the conflicting traffic when time does not allow the above procedure to be carried out.

Under RAS, the following conditions apply:

a The service may be requested under any flight rules or meteorological conditions
b The pilots are expected to accept vectors or level allocations by ATC which require flight in IMC. (**Pilots unqualified to fly IMC should accept an RAS only where ATC advice allows VMC flight**)
c When a pilot chooses not to comply with advised avoiding action, he/she must inform the controller, and is then responsible for any subsequent avoiding action necessary
d The controller must be advised by the pilot of any changes in heading or level
e Conflicting traffic information will be passed until the confliction has been resolved
f Although no legal requirements outside controlled airspace demand compliance with ATC instructions, the pilot will be assumed to comply unless he states otherwise.

Radar Information Service

The RIS is an air traffic service whereby the pilot is wholly responsible for the aircraft separation between his/her aircraft and other aircraft, regardless of the traffic information passed to him/her by the controller. The controller will inform the pilot of the bearing, distance and, if known, the level of other conflicting traffic, but will not offer avoidance action. The following conditions apply when a pilot is using the RIS.

a The service may be requested under any flight rules or meteorological conditions
b After initial warning, the controller will only update details of any

conflicting traffic at the pilot's request, or if the controller considers that the conflicting traffic remains a definite hazard

c The pilot must inform the controller before changing level, level band or route

d RIS may be offered when the provision of RAS is impracticable

e Requests for an RIS to be upgraded to a RAS will be accepted subject to the controller's workload; standard separation will be applied as soon as practicable. If an RAS cannot be provided, the controller will continue to offer an RIS.

ESTABLISHING A SERVICE An agreement must be made between the pilot and the controller when establishing a radar service, and the pilot must state whether he/she requires an RAS or RIS. The controller will attempt to identify the aircraft if he/she is able to offer a service, and once positive identification has been confirmed the controller will state the type of service he/she is about to provide. **The pilot must not assume that an RAS or RIS has been confirmed by the identification procedure, and must await positive confirmation that a radar service is being provided by the controller**.

Note: If the pilot fails to specify the type of service required, before providing any service, the controller will ask the pilot which type of service is required.

Radar Control Service

Radar Control Service gives positive control of aircraft in controlled airspace, ensuring that radar separation is given to aircraft arriving, departing and en route. This is accomplished by monitoring en route and approach aircraft and, where necessary, giving radar vectoring to such aircraft. Being part of the control service, it gives assistance to pilots in navigation and in crossing controlled airspace. Apart from maintaining separation of aircraft from each other, it also supplies a proximity hazard warning to obstacles which may affect the aircraft. It may also supply information on observed weather conditions and give assistance to aircraft in distress.

SEPARATION STANDARDS The standard horizontal separation is 5 nm (9.3 km), but this may be reduced to 3 nm (5.6 km) within 40 nm of the radar head and below FL 245 when official approval is given for the procedure. When official approval of SSR only is given, the distance will be increased to 8 nm (14.8 km).

Terrain Clearance

The clearance levels given by controllers to IFR flights in receipt of Radar Control Service and flights receiving RAS will ensure that the minimum terrain clearance will be at least one of the following:

1 Excluding the Final and Intermediate Approach Area, the terrain clearance within 30 nm of the radar antenna is 1,000 ft above any fixed obstacle which is closer than 5 nm to the aircraft. It also gives 1,000 ft clearance above any fixed obstacle within the area 15 nm ahead of and 20 degrees either side of the aircraft's track. These distances can be reduced to 3 nm and 10 nm respectively with CAA approval.

2 On Airways and Advisory Routes beyond the 30 nm from the radar antenna, clearance is 1,000 ft above any fixed obstacle within 15 nm of the centreline, otherwise 1,000 ft above any fixed obstacle within 30 nm of the aircraft.

(NOTE: In sections of Airways where the base is defined as a Flight Level, the lowest usable level will provide a minimum terrain clearance of 1,500 ft.)

3 Aircraft receiving RIS, and those operating under Special VFR or VFR accepting radar vectors, will not be assigned levels by the radar controllers, and the pilots are responsible for their own terrain clearance.

WEATHER AVOIDANCE ATC will vector all aircraft operating in Controlled Airspace to remain at least within 2 nm inside of the Controlled Airspace boundary. When the weather conditions are bad enough to affect the planned route of an aircraft in Controlled Airspace, the following conditions apply:

1 If the ATC radar detects bad weather, the pilot will be advised by the controller. However, the pilot is responsible for accepting any detour into uncontrolled airspace.

2 If the pilot detects bad weather conditions on the aircraft's radar when in Controlled Airspace, the pilot must obtain clearance from the radar controller before leaving Controlled Airspace, and must also request permission to rejoin when the bad weather conditions have cleared.

SECONDARY SURVEILLANCE RADAR OPERATION SSR operation in the UK is to be used for aircraft detailed in Schedule 5 of the ANO in airspace as laid out in the following scale:

1	The whole of the UK Airspace at and above FL100	Mode A and Mode C with altitude reporting
2	UK Controlled Airspace notified for the purposes of Schedule 5, sub-para 2 (1) (a) below FL100 when operating under IFR	
3	The Scottish TMA. Between 6,000 ft ALT and FL100	Mode A

The above requirements apply to all aircraft except gliders and aircraft below FL100 in Controlled Airspace notified for the purposes of Schedule 5, receiving an approved crossing service.

Basically, SSR is a radar system consisting of an interrogator (a transmitter) and a transponder. The interrogator sends out a transmission which is received by the aircraft, and a return signal is then decoded and supplies information to the ATC. A preselected code is put into the transponder which identifies the aircraft, giving ATC the aircraft's position and FL, thereby helping to maintain safe separation.

In airspace where mandatory use of transponders is not required, pilots of suitably equipped aircraft should comply with the regulations of the conspicuity code, with the exception of those flying below 3,000 ft AGL remaining within an aerodrome traffic pattern.

When pilots are not required to use the Special Purpose Codes, they shall conform to the following requirements:

1 When an ATS unit has assigned a specific code to an aircraft, the pilot should maintain that code setting unless otherwise instructed.
2 The pilot should select or reselect codes, or switch off equipment when airborne, only when instructed to do so by an ATS Unit.
3 The pilot should read back code settings to acknowledge ATC instructions.
4 The pilot should select Mode C simultaneously with Mode A unless instructed by an ATS Unit.
5 When reporting levels under routine procedures or when requested by ATC, the pilot should state the current altimeter reading to the nearest 100 ft. This is to assist in the verification of Mode C data transmitted by the aircraft. Once verified, if there is a difference of more than 200 ft between the level readout and the reported level, the pilot will normally be instructed to switch off Mode C. If independent switching of Mode C is not possible, the pilot will be instructed to select Code 0000 to indicate a transponder malfunction.

SPECIAL PURPOSE SSR CODES Some codes have been reserved internationally for special purposes. These codes should be operated with Mode C, and should be selected under the following conditions:

a Code 7700 – Indication of an emergency condition but if an ATS code is already under transmission, that code will normally be retained
b Code 7600 – Indication of a radio failure
c Code 7500 – Indicates unlawful interference with the planned operation of a flight, (i.e. hijacking), unless circumstances warrant the use of Code 7700

d Code 2000 – When entering United Kingdom Airspace from an adjacent region where the operation of transponders has not been required

e Code 7007 – Allocated to aircraft engaged on airborne observation flights under the terms of the Treaty of Open Skies. Flight Priority Category B status has been granted for such flights, and details will be published in NOTAMS.

CONSPICUITY CODE Unless otherwise directed by an ATS Unit, Code 7000 and Mode C should be selected by pilots except:

a When receiving a service from an ATS unit or Air Defence Unit which requires a different setting

b When circumstances require the use of one of the Special Purpose Codes.

NOTE: Code 7000 is in close proximity to the Special Purpose Codes, and therefore pilots are warned of the need for caution when selecting Code 7000.

PARACHUTE DROPPING Pilots of aircraft equipped with transponders should select Code 0033, together with Mode C, five minutes before the drop commences and until the parachutists are estimated to be on the ground. The exceptions to this is when a discrete code has previously been assigned.

Transponder Failure

In the event of transponder failure, pilots are required to carry out certain actions.

BEFORE TAKE-OFF If the transponder fails before the aircraft intends to depart and it cannot be repaired at the departure aerodrome, the pilot shall proceed to the nearest suitable aerodrome where repairs can be carried out. ATS should be informed by the pilot, if possible before the submission of a flight plan. In this situation ATC may delay or alter the take-off clearance, dependent upon the existing and anticipated traffic situation. Flight Level clearances and intended routes may be modified to comply with traffic situations.

In the case of transponder unserviceability, the correct letter should be inserted in item 10 of the flight plan as laid out in the ICAO Doc 4444, Appendix 2, dependent upon either total failure or partial failure of the equipment.

AFTER TAKE-OFF ATS Units will try to provide a continuation of the flight according to the flight plan if the transponder fails after take-off or en route. However, depending upon how soon after take-off, or the traffic situation, when the failure takes place, the aircraft may be required to return to the departure aerodrome or to land at a suitable aerodrome acceptable to ATC.

If after landing the transponder cannot be repaired under any circumstances, the pilot must carry out the requirements of 'failure before take-off'.

Lower Airspace Radar Service

Within 30 nm of each participating ATS unit, radar/radio cover is made available to all aircraft outside controlled airspace between the surface and FL95 inclusive. The hours of operation depend upon each ATS unit. However, if it is not classed as a 24-hour operation, flying may make the ATS unit operational outside the standard published hours. Pilots are therefore advised to attempt to make contact with the ATS unit by making three consecutive calls, after which it can be considered unavailable.

Controller's advice given under the RAS is designed to enable certain minimum separation criteria. These criteria are:

1 A minimum horizontal separation of 3 nm between all identified aircraft working the same unit.
2 A minimum horizontal separation of 5 nm between identified aircraft and other observed aircraft, unless it is known that a 1,000 ft vertical separation exists.

It is not mandatory for any pilot to heed the advisory avoiding instructions, but if the pilot decides not to, he/she should inform the controller immediately. The pilot must then assume the responsibility for initiating subsequent avoiding action.

The controller must be warned before any changes of heading or flight level/altitude when under a radar service, and vertical separation should be in accordance with the Quadrantal Rule.

RAC 3-8-3-1 (14 Nov 91)

As pilots may cross several different radar areas, aircraft will, whenever possible, be handed over from controller to controller in areas of overlapping radar. When handing over is not possible, pilots will be informed of their position and advised which unit to call for further assistance.

Military Middle Airspace Radar Service

This is a service similar to the LARS, but it covers airspace between FL100 and FL240, except for Brize Norton, which has an upper limit of FL150. This service is available to all aircraft flying outside Controlled Airspace in the UK FIR, except for flight:

1 Along advisory routes.
2 Within the Northern Radar Advisory Service Area (NRASA).
3 Within the Sumburgh FISA.

RAC 3-8-8-1 (14 Nov 91)

CAUTION: Owing to a unit's capacity being exceeded in the LARS and MARS, controllers may not be aware of some aircraft, so pilots should maintain a careful look-out at all times.

Military Mandatory Radar Service Area

Procedures are laid down for aircraft above FL245 and operating within the MRSAs and outside the MRSAs. Within these areas aircraft are under the control of different ATS units to ensure separation between aircraft. Aircraft that are flying off the promulgated ATS routes are normally vectored clear of aircraft flying on the routes.

Only if there is an indication that an aircraft is lost or experiencing radio failure will traffic information and/or avoiding action be given.

When an aircraft is flying above FL245 and outside the MRSAs, traffic information is given when practically possible. However, avoiding action will be given only if requested by the pilot or if the controller deems it necessary.

Chapter 8
Flight Procedures

This chapter contains the information for en route flying, covering the departure, climb, en route and approach procedures. The procedures required when flying in circuit and in the immediate vicinity of an aerodrome with the intention of landing at the aerodrome of departure are related to in Chapter 5, Aerodromes and Procedures. The ATC organisation worldwide is designed to control the flow and management of aircraft with the intention of preventing accidents and mid-air collisions. To achieve this, a flight separation standard is used with the use of flight plans and specifically detailed rules.

Flight Plans

A flight plan is a standard form requiring specific details relating to the flight, the aircraft and the equipment on board. It also contains information about the number of persons on board, destination, route to be followed, etc. The manner in which the form is to be completed is covered in AIC 134/1993 (Yellow 118).

There are times when a flight plan *may* be filed by the pilot, when the pilot *must* file a flight plan, and also when a pilot is *advised* to file a flight plan.

1 A flight plan *may* be filed for any flight.

2 A flight plan *must* be filed for:

all flights within:

a Class A Airspace
b Any Controlled Airspace in IMC or at night, excluding Special Visual Flight Rules (SVFR)
c Any Controlled Airspace in VMC if the flight is to be conducted in accordance with IFR
d Class B, C and D Control Zones/Control Areas, irrespective of flight rules

e The London and Scottish Upper Flight Information Region

any flight:

f From an aerodrome of departure in the United Kingdom, being a flight having a destination more than 40 km from the aerodrome of departure and in which the aircraft's MTWA exceeds 5,700 kg

g To or from the United Kingdom which will cross the UK FIR boundary.

3 A flight plan is *advised* to be filed if the flight involves flying over the sea, more than 10 miles from the UK coast, or flying over sparsely populated areas where search and rescue operations would be difficult.

Flight Separation

Flight separation is made both in the vertical and horizontal planes. In the vertical plane, a minimum separation is 1,000 ft between two aircraft on reciprocal tracks up to FL290, and 2,000 ft above FL290. There is also provision for a vertical separation between a holding aircraft and an en route aircraft when the en route aircraft is 5 min flying time away. An aircraft at a cruising level normally has priority over aircraft ascending or descending, and, when two aircraft are at the same cruising level, priority is normally given to the preceding aircraft.

Horizontal separation is further defined in three categories; lateral separation, longitudinal separation and radar separation.

1 Lateral separation

Lateral separation is subdivided into track separation and geographical separation.

Track Separation requires minimum separation dependent upon the navigational aid or method employed:

VOR Tracks require a separation of at least 15 degrees and at a distance of 15 nm from the facility.

NDB Tracks require a separation of at least 30 degrees and at a distance of 15 nm or more from the facility

DR require the tracks to be diverging by at least 45 degrees and at a distance of 15 nm or more from the point of intersection of the tracks, this point being determined either visually or by reference to a navigational aid.

Geographical Separation must be positively indicated by position reports over various geographical locations determined by visual reference or a navigation aid.

2 Longitudinal separation

Longitudinal separation is the separation that aircraft have from each other over a specific geographical position. This allows the time interval between aircraft to be at a minimum, or greater than the specified minimum allowed. This caters for aircraft at the same level, on the same track, for aircraft climbing or descending and aircraft on their departure take-off climb-out. Separation can therefore be based on time or distance, as follows:

BASED ON TIME

1. For aircraft in rapid fixing areas on the same track, on the same cruising level, crossing tracks and climbing or descending there is a 10 min separation time.
2. For aircraft at the same cruising level on the same track, crossing tracks, climbing or descending, the normal separation time is 15 min.
3. For aircraft at the same cruising level on the same track, when the preceding aircraft is at least 20 kt (TAS) faster, the minimum separation is 5 min, and if it is at least 40 kt (TAS) faster the minimum separation is 2 min.
4. For aircraft climbing or descending when the level change is commenced within 10 min of the time that the second aircraft has reported over an exact reporting point, the minimum separation is 5 min (as shown in Fig. 8-1).

BASED ON DISTANCE When two aircraft are on the same track or crossing tracks to or within the area of a DME, the separation must be of at least 20 nm. If two aircraft are on the same track and the leading aircraft is 20 kt faster, the separation distance must be at least 10 nm. When crossing tracks with the first aircraft 20 kt faster, the second aircraft must have a separation of at least 10 nm. When climbing or descending by an aircraft, a separation of 10 nm must exist between the two aircraft.

3 Radar separation

Radar separation allows for a minimum of 5 nm in the horizontal plane between aircraft, but under CAA approval this may be reduced to 3 nm when the aircraft are within 40 nm of the radar head and below FL245. If only SSR is available, the distance will be increased to 8 nm.

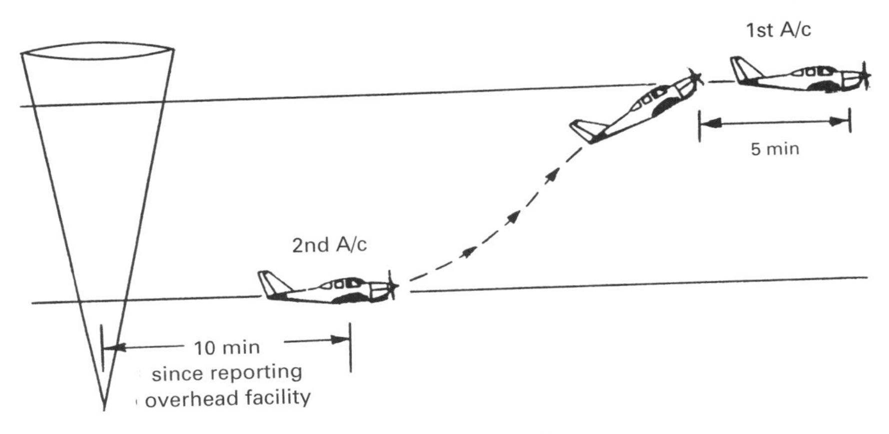

Fig 8–1 Time Separation

Departure Procedures

Departure from aerodromes are not always straightforward, as in many circumstances there are obstacles to be avoided, special minimum climb requirements, and also noise abatement procedures to be carried out. Therefore departure procedures are normally one of the following:

a Instrument departure procedure
b Omnidirectional departure procedure.

Instrument departure procedures are published in order to ensure acceptable terrain clearance, and obstacle clearance criteria are observed during the departure phase. This may be achieved by using any one or combination of the following:

a specific sectors to be avoided
b specific routes to be followed
c minimum net climb gradients to be achieved.

The instrument departure procedures are achieved using navigational aids on the outbound track, which may be a straight departure or a turning departure. Detailed criteria for both the instrument departure procedure and the omnidirectional departure procedure are laid down in *PANS-OPS (DOC 8168) VOLUME 1*. For the purpose of the air law syllabus the required information is covered in this chapter, but study of *PANS-OPS (DOC 8168) VOLUME 1* will enhance the pilot's knowledge of procedures.

To ensure that correct separation limits are observed, flights intending to operate on a route subject to flow regulations, and departing from aerodromes within the United Kingdom and the Republic of Ireland, must first obtain departure approval from the London Flow Management Unit either directly, by:

a the AFTN; or
b SITA, using a Slot Request Message (SRQ),

or via the National Air Traffic Flow Management Service (NATFMS) Reception Position by means of the telephone, fax or telex.

Before take-off the altimeter should be checked and set. The *aerodrome apron* is the designated location for this check within the United Kingdom. The desired setting for take-off (e.g. QFE or QNH) is at the discretion of the pilot, but one altimeter should be selected to the latest QNH setting for that aerodrome.

Aerodrome ATC clearance must be obtained before take-off, and it will specify some or all of the following information as necessary:

a Direction of take-off and turn after take-off
b Track to be made good before proceeding on desired heading
c Level to maintain before continuing climb to assigned cruising level, time, point and/or rate of which level change shall be made
d Any other manoeuvre necessary to maintain separation as appropriate.

While on the aerodrome QNH, the vertical position of the aircraft will be expressed in terms of altitude during the climb to and while at Transition Altitude. Once clearance has been given to climb above the Transition Altitude, the vertical position will be given in Flight Levels once that climb has commenced, provided the aircraft is not more than 2,000 ft below the Transition Altitude. Any difference to this procedure will only be under a specific request by the ATC.

When the clearance to climb above Transition Altitude has been given, the Flight Level position reports must be given with the altimeter set to 1013.2 mb. Except for selected airspaces within the UK, as notified in the RAC 2-2 of the *UK Air Pilot*, the Transition Altitude within the United Kingdom is 3,000 ft.

To check for Terrain Clearance, the most appropriate lowest forecast regional QNH value should be used.

Terrain clearance

The base of an Airway is at least 1,000 ft above any fixed obstacle within 15 nm of the centreline. However, an absolute minimum altitude applies when the lower limit of an Airway section is defined as a Flight Level. The absolute minimum altitude clearance is 1,500 ft above any obstacle within 15 nm of the Airway centreline. Therefore the lowest usable level will always be 500 ft above the Airway base.

Obstacle clearance

Obstacles are classified into aerodrome obstacles and en route obstacles. Those obstacles within 4 nm of the Aerodrome Reference Point (ARP) are aerodrome obstacles, and those obstacles located more than the 4 nm from the ARP are en route obstacles. An Air Navigational Obstacle is defined as any buildings or works, including waste heaps, which attain or exceed a height of 300 ft AGL.

AERODROME OBSTACLES A low-intensity steady red light is used to light fixed obstacles of 45 m or less in height, width and length. For obstacles of 45 m or greater, up to but not exceeding 150 m in height, medium-intensity flashing red lights are used. High-intensity flashing white lights are used for all obstacles greater than 150 m in height.

EN ROUTE OBSTACLES Any obstacles that can be considered, by virtue of their nature or location, significant hazards to air navigation and are 492 ft AGL (150 m) or less in height, are lit by medium-intensity red flashing lights by night. Obstacles en route that *exceed* 492 ft AGL (150 m) in height are lit day and night by high-intensity flashing white lights.

Position reports

Along the routes within the controlled airspace certain points are allocated as position reports where the time, position and flight level of the aircraft has to be reported to the ATCU. Position reporting points are annotated on the charts as solid black triangles (▲) which are Designated Reporting Points, or outlined triangles (△) which are 'on request' reporting points. 'On request' reporting points require a report to be made only when requested by the controlling authority. The ETA at the next reporting point is normally also transmitted in the report.

Airborne Procedures

There may be times when an aircraft requires to join or leave an Airway en route instead of at a TCA. Therefore once clearance has been given to an aircraft to leave or join an Airway at a particular point, the flight should be flown to cross the Airway actual boundary as close to that point as possible.

Adherence to IFR procedures is required when flying Airways, regardless of weather conditions. However, when climbing or descending in Airways when no radar cover is available, ATC may clear a VMC climb or descent subject to the following conditions:

a VMC during daylight only
b Subject to the pilot's agreement
c Separation responsibility rests with the pilot
d Essential traffic information will be given.

EN ROUTE HOLDING En route holding must be carried out on tracks parallel to the Airway centreline, turning right at the reporting point. The only exception to this is under ATC instructions. Holding patterns should be adjusted accordingly when pilots are given a specific time to leave the reporting point.

Pilots are required to report as follows:

a The time and level of reaching a specific holding point to which cleared
b When leaving a holding point
c When vacating a previous assigned level for a new assigned level.

JOINING AIRWAYS A flight plan must be filed either before departure or when airborne, and a request for joining clearance must be made by the pilot at least 10 min flying time before the intended joining point.

Clearance to join the Airway must be initiated with a call to ATC, giving identification, requesting joining clearance of which Airway, and at which point. When requested by ATC, the following details should be passed:

a Identification
b Aircraft type
c Position and heading
d Level and flight conditions
e Departure aerodrome
f Estimated time at entry point
g Route and point of first intended landing
h True airspeed
i Desired level on Airway (if different from above).

Crossing airways   An aircraft may cross an Airway without ATC clearance if it flies at right angles across the base of an en route section where the lower limit is defined as a Flight Level.

To cross an Airway at any other level, the pilot must file a flight plan either before departure or when airborne, and must also request clearance to cross at least 10 min flying time before the intended point of crossing.

The initial call should be made as with joining an Airway, and, when instructed by ATC, the following information should be passed:

a Identification
b Aircraft type
c Position and heading
d Level and flight conditions
e Position of crossing
f Requested crossing level
g Estimated time of crossing.

Gliders may cross an Airway, except a Purple Airway, provided it is done in the most expeditious manner and at right angles to the Airway centreline. It must be carried out by day in VMC. This VMC minima is to be 8 km visibility, 1,500 horizontal and 1,000 ft vertical distance from clouds.

Approach procedures

Arrival procedures for VFR flights

When an aircraft under VFR is 15 nm or 5 min flying time from the ATZ boundary, whichever is the greater, RTF contact should be made with the Approach Control service. If the Approach frequency is not known to the aircraft pilot, initial contact may be made on the Aerodrome Control frequency. ATC will pass the landing information and any pertinent known traffic to VFR flights to ensure that separation from IFR and other VFR traffic is maintained.

VFR flights will be supplied with information by ATC enabling them to fit into the landing sequence when radar sequencing of IFR flights are in progress. Approach Control will inform the pilot of the frequency of, and when, to change to Aerodrome Control.

Approach Control service

The Approach Control service is available at aerodromes inside and outside of Controlled Airspace. However, unless they are within the Aerodrome Traffic Zone, pilots flying IMC are not legally required to comply with

Approach Control instructions if they are outside Controlled Airspace. No legal requirements necessitate pilots in Uncontrolled Airspace to report their presence. It is because Approach Control at aerodromes in Uncontrolled Airspace cannot be certain of aircraft separations that all information passed to the pilot must be regarded as advisory only.

To improve the safety aspect, pilots contacting Approach Control at aerodromes in Uncontrolled Airspace are strongly recommended either to:

a Avoid IFR flying below 3,000 ft and within 10 nm of an aerodrome having Approach Control, or
b Contact Approach Control at least 10 min flying time away and comply with any instructions given if it is necessary to fly IFR in Uncontrolled Airspace near an aerodrome.

Approach Control responsibility

CONTROLLED AIRSPACE AERODROMES Standard separation is provided to inbound aircraft on IFR flights that are released by ATCC or Zone Control, and to aircraft under its jurisdiction from the FIR within Controlled Airspace until they are transferred to Aerodrome Control.

Standard separation is also provided to outbound aircraft by Approach Control once handed over from Aerodrome Control until handed over to the ATCC.

UNCONTROLLED AIRSPACE AERODROMES *Arriving aircraft* in Uncontrolled Airspace, although subject to receipt of advisory information, will be provided with separation by Approach Control on being released by ATCC or by placing themselves under Approach Control jurisdiction until they are transferred to Aerodrome Control.
Departing aircraft will be provided with separation by Approach Control once transferred by Aerodrome Control, until:

a They are transferred to ATCC, or
b They state they no longer wished to be controlled, or
c They are more than 10 min flying time away from the aerodrome,

whichever is the sooner.
Transit aircraft first place themselves under control of Approach Control until they state they no longer wish to be controlled or they are clear of the approach pattern.

Arriving aircraft

Pilots flying in Controlled Airspace should contact Approach Control when ATCC instruct them to do so, while pilots flying in Uncontrolled Airspace in IMC should call Approach Control at least 10 min before their aerodrome ETA. Once contact has been made with Approach Control, the pilot will be given the following information.

a Runway in use
b Current meteorological information, to include:
  1 Surface wind direction (in degrees magnetic) and speed
  2 Visibility
  3 Present weather
  4 Significant cloud amount and height of base
  5 QFE or QNH (with height of aerodrome)
  6 Any other relevant information (gusts, icing, etc.)
  7 Runway Visual Range.

NOTE: When aircraft are below cloud in VMC until the landing, information passed may be given as surface wind and speed with the QFE/QNH.

c Current runway surface conditions where appropriate
d Any change in operational status of visual and non-visual aids essential for approach and landing.

HOLDING A pilot may be required to hold, dependent upon the traffic situation. A time will be given to the pilot when to leave the hold and to commence an approach. If the pilot is unable to comply with the holding instructions, on request by the pilot an alternative may be permitted.

INSTRUMENT APPROACHES Although in exceptional circumstances Approach Control will supply certain information, pilots are expected to know and be familiar with (through charts and published information), the correct instrument approach procedure for the aerodrome intended for landing. In exceptional circumstances Approach Control will supply:

a The navigational aid concerned
b The initial approach level
c Outbound track, time in minutes, and level instructions
d Direction of procedure turn and level instructions
e Final approach Track and level instructions
f Obstacle clearance height
g Missed-approach procedure.

A pilot may be given, upon request, permission to complete the approach visually when he/she can see the ground before the approach procedure is completed, otherwise the entire procedure must be carried out. Permission to complete a visual approach will only be given when:

a The pilot can maintain visual reference to the surface, and
b The reported cloud ceiling is not below the initial approach level, or the pilot reports that the visual approach can be continued and a landing accomplished due to the visibility.

An instrument approach may have one or more of the following segments (Fig. 8-2):

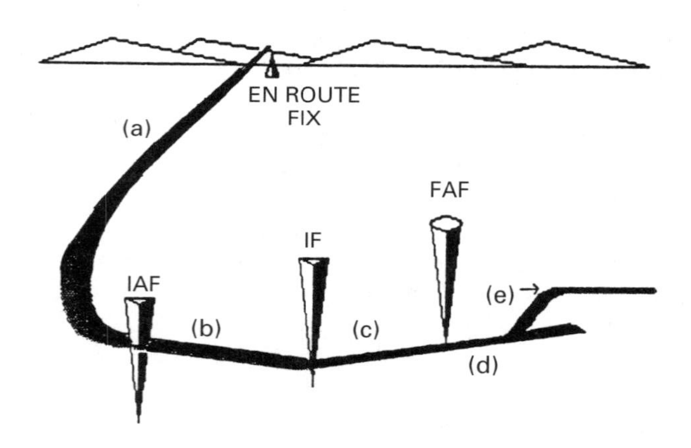

Fig 8–2 Instrument Approach Segments

a A feeder segment (en route fix)
b An initial approach segment (IAF)
c An intermediate approach segment (IF)
d A final approach segment (FAF)
e A missed approach segment.

Fix

The segments of an instrument approach normally begin or end at designated fixes. A fix is a definite position of an aircraft that is determined visually by the intersection of:

a ADF bearings and VOR radials
b VOR radials and DME distance indications
c ADF bearings.

Aircraft can also be vectored over a fix by ground radar personnel.

Feeder segment

A feeder segment is not required for all instrument approaches. When required, the feeder segment is used to designate the course and distance from an en route fix on a VOR airway to the initial approach fix (IAF) for the landing runway.

The feeder segment provides the exact course and altitude that must be flown to arrive over the initial approach fix for the instrument approach to be flown.

Initial approach segment

In the initial approach segment, an aircraft departs the IAF and manoeuvres to enter the intermediate segment. An initial approach segment can be made along an arc, radial, course, heading, radar vector, or any combination of these. Procedure turns, holding pattern descents, and high-altitude teardrop penetrations are also initial segments.

Intermediate approach segment

The intermediate approach segment begins at the intermediate fix (IF) and ends at the final approach fix (FAF). The intermediate approach segment transitions an aircraft from the initial approach fix to the final approach segment. In the intermediate approach segment, the aircraft configuration, speed, and positioning adjustments are made for the final approach segment.

Final approach segment

The final approach segment is the part of the instrument approach in which the final alignment and descent for the landing are accomplished. The final approach segment begins at the FAF and ends at the missed approach point. A final approach can be made to a runway for a straight-in landing, or to an airport for a circling approach.

Missed approach segment

A missed approach segment begins at the missed approach point, and provides obstruction clearance and course guidance to a fix for holding or returning to a VOR airway. The missed approach point can be the intersection point of an ILS glidepath with a decision height (DH) or a minimum descent altitude (MDA). It can also be a navigational facility, a fix, or a specific distance from the final approach fix.

CHAPTER 9

SIGNALS AND COMMUNICATION

Four basic types of communication services are used in aviation. These are:

a The Aeronautical Radio Navigation Service
b The Aeronautical Mobile Service
c The Aeronautical Fixed Service
d The Aeronautical Broadcast Service.

The Aeronautical Radio Navigation Service

Navigational radio aids used in the United Kingdom are:

a MF non-directional beacon (NDB)
b VHF direction finding station (VDF)
c Precision approach radar (PAR), which is available at certain military aerodromes
d Approach radar (RAD)
e Instrument landing system (ILS)
f Microwave landing system (MLS)
g VHF omni-directional radio range (VOR)
h Distance measuring equipment (DME).

MF non-directional beacon

These beacons, used in conjunction with the aircraft's ADF, are used for en route and airfield approach facilities. Although they have been increasingly replaced by the VOR facilities, the NDBs are still used widely throughout the world, and may continue to be used for several years to come. The stated range of the NDBs used in the UK are based on a daytime protection ratio between the wanted and unwanted signals, which is achieved by geographical separation and/or power adjustment of various facilities. This limits the bearing errors at a distance to plus or minus 5 degrees or less. The bearing errors will increase at ranges greater than those stated, and also through adverse propagation conditions or especially at night. Therefore at night it is important for the ADF to be kept correctly tuned.

The use of NDBs as an approach aid is limited to the ATC's notified hours of service. However, most locator beacons continue to transmit a usable signal outside these hours for navigational purposes only.

The NDBs transmit a Morse identification signal which can be one of two types. A modulated continuous-wave-type of modulation is used in the United Kingdom, but many other countries use an interrupted continuous-wave modulation which requires a beat frequency oscillator (BFO) or a tone generator in the ADF receiver. It is therefore essential for pilots to be aware of the type of emission to expect, and to preselect the ADF receiver accordingly.

VHF direction finding station

Classification of the VDF bearings are as follows:

1 Class A; accurate to within plus or minus 2 degrees.
2 Class B; accurate to within plus or minus 5 degrees.
3 Class C; accurate to within plus or minus 10 degrees.

When satisfactory conditions exist, VDF bearing information will be given, although a Class B bearing is normally the best available.

Precision approach radar

PAR is found at major civil aerodromes and most military ones. With PAR the controller provides the pilot with highly accurate navigation guidance in azimuth and elevation. This is accomplished by two independent radar units which project a beam on two radar displays, one in azimuth and one in elevation. The approach track beam is approximately half a degree wide and two degrees deep, and scans ten degrees either side of the centreline. The approach elevation beam is two degrees wide and half a degree deep, scanning through a seven-degree arc in the vertical plane. Both have range markers at one-nautical-mile intervals.

Pilots are initially advised of headings to fly to direct them on to the extended centreline of the landing runway. They are then informed about glidepath interception approximately 15 to 30 sec before it occurs, and are advised to anticipate descent at interception. The range from touchdown is passed to the pilot at least once every mile. The pilot is advised as being either above or below glidepath and right or left of the centreline. The controller terminates the information on completion of the approach, when the aircraft is over the threshold for landing.

Approach radar

This is a surveillance radar approach, and the controller gives only azimuth approach information to pilots. Headings are given to place pilots on the extended centreline of the landing runway. Although only azimuth information is passed to pilots, the controller will tell them when to begin their descent down to the MDA. The controller will, at a pilot's request, pass recommended altitudes at each mile down to the last mile that is at or above the MDA.

Instrument landing system

The ILS is a system designed for making a precision IFR approach on to the runway at an aerodrome. The system consists of a localiser and glidepath aerials situated at the aerodrome, and ILS receiving equipment contained in the aircraft. A middle marker is situated approximately 1,500 m from the end of the runway along the extended centreline, and an outer marker is situated approximately 8 km from the end of the runway along the extended centreline.

THE LOCALISER The localiser transmitter supplies pilots with azimuth direction information either side of the runway centreline. The coverage area of the localiser is 35 degrees either side of the front course line up to 17 nm and 10 degrees either side of the front course line up to 25 nm (see Fig. 9-1 page 106). Owing to interference, the localiser transmission is only protected to a range of 25 nm and to an altitude of 6,250 ft, although the accuracy is only checked to 10 nm.

For steep-angle glide paths, the localiser's coverage area is reduced to 35 degrees either side of the front course line up to 10 nm, and 10 degrees either side up to 18 nm.

THE GLIDEPATH The glidepath transmitter is situated approximately 150 m to one side of the runway and displaced approximately 250 to 300 m from the landing threshold. Two aerials are used, giving upper and lower lobes which overlap to provide a glidepath equisignal for true glidepath information. The ILS glidepath coverage is 8 degrees either side of the extended runway centreline to 10 nm in the azimuth angle. Vertical coverage and limitations are more complicated, and are therefore laid down in an AIC pink leaflet.

NOTE: *There are pink leaflets in the AICs covering ILS airborne equipment with reduced channel spacing, and also the use of ILS facilities in the United Kingdom. These should be read and understood by all pilots.*

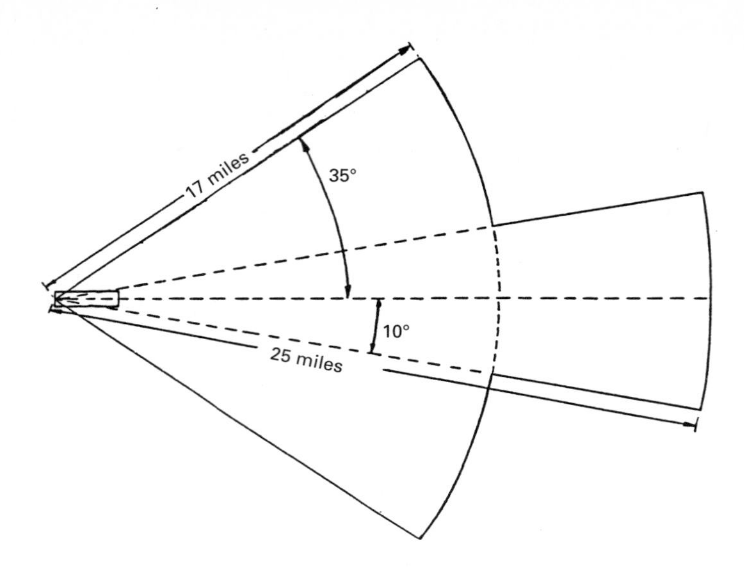

Fig 9–1 ILS Localiser Range

All ILS systems vary in their performance capability, and they are therefore categorised as follows:

a Category I — Guidance information is provided from the coverage point down to a glidepath decision height of 200 ft above the ILS reference point.

b Category II — Guidance information is provided from the coverage point down to a glidepath decision height of 50 ft above the ILS reference point.

c Category III — Guidance information is provided from the coverage point down to, and along the surface of, the runway.

FREQUENCIES The localiser channels are on the VHF frequency band of 108.10 to 111.9 kHz (odd decimals), with a channel spacing of 50 kHz.

The glidepath channels are on the UHF frequency band of 329.3 to 335.0 mHz, with a channel spacing of 150 kHz.

Microwave landing system

The MLS is similar to the ILS, but the azimuth coverage area is plus or minus 40 degrees about the nominal courseline and has a range of 20 nm unless otherwise stated.

VHF omni-directional radio range

The VOR navigation aid is often used with DME to determine the position of the aircraft along a radio signal radial from the VOR transmitter. The VOR transmits a signal on 360 degree radial angles which is picked up by a receiver in the aircraft. This indicates the direction of the aircraft from the VOR transmitter, and when paired with the DME signal it can also give a range along that radial, pinpointing the aircraft's position. As it is a VHF signal from ground to air, its normal maximum range is approximately 200 nm, although it has been known to pick up signals beyond that range.

The VOR's accuracy is affected beyond the 200-mile range, and owing to small receiver equipment inaccuracies there may be inaccuracies at closer ranges. This can be noticed when flying between two VOR transmitter beacons less than 200 miles apart. The navigational reading may indicate that one or the other does not place the aircraft on track correctly. This phenomenon is known as 'bending' or 'scalloping'. VOR transmitters are monitored to a 1- degree accuracy, and if the inaccuracy is greater than 1- degree the VOR at fault will automatically switch off and change to a standby transmitter. In this case no identification signal is radiated until the changeover is completed. The VOR bearing information is protected by a published value referred to as the Designated Operational Coverage, and is listed in the COM 2 section of the *UK Air Pilot*.

Distance measuring equipment

The DME is designed to give the distance from a known locator, which is normally situated at an aerodrome. It uses the UHF frequency bands of 962 mHz to 1213 mHz, giving a slant range from the transmitter. This means that an aircraft flying directly overhead the transmitter at approximately one mile high would have a DME readout of 1 mile, being 1 mile directly overhead. The DME is normally used in conjunction with VOR or TACAN to give an accurate position from a transmitting station.

The Aeronautical Mobile Service

The Aeronautical Mobile Service provides a radio communication service for use between the air and ground services, and for air-to-air use between different aircraft for both normal communication traffic and also for emergencies. The language of the service is that as used by the country of travel, although English is the most widely used language throughout the world, to reduce errors in translation and to provide a continuity of service.

To ensure that communication between air traffic services and aircraft is efficiently maintained, R/T transmissions must be kept to a minimum to avoid congestion. Typical examples of radio congestion are as follows:

a Failing to comply with standard reporting procedures
b Failing to listen out before transmission
c Lack of radio discipline
d Requests for information covered already by met broadcasts
e When requested, failing to restrict signals
f Use of unauthorised abbreviations
g Leaving the transmission switch open.

During the course of a flight, a change in radio frequency may be needed, either for ATC to pass the pilot on to another controller en route, or on approach and landing. It is therefore essential that pilots read back to the ATCO the frequency to which they are changing when told of frequency changes by the ATCO, to ensure that safe and efficient handover has occurred, and not a radio failure. The ATCO will assume radio failure has occurred if this procedure is not complied with.

Aeronautical emergency channels are set aside for emergency communications, to provide:

a Clear channels for aircraft in distress
b Communication between search aircraft
c Emergency communication between an aircraft and the ground
d A means of communication with ocean weather vessel stations when no other channel is available
e Means of communication when all other channels fail.

Channels for international aeronautical emergency and distress frequencies in use on a world-wide basis are:

a 121.5 mHz (VHF)
b 243.0 mHz (UHF)
c 500 kHz (HF)
d 2182 kHz (HF).

Use of these frequencies is described in Chapter 11, along with all procedures for emergency calls and radio failure procedures.

SELCAL

A selective calling system (SELCAL) used on HF and VHF frequencies is used to allow pilots the chance to discontinue direct monitoring of the frequency in use, especially if that frequency (mainly HF frequencies) has a large amount of static interference. It allows a ground station to alert pilots that they are being called for radio transmission purposes. Each aircraft is assigned a tone code, which when activated, is received through a cockpit call system in the form of a light and/or chime signal. When pilots prepare flight plans they should enter this SELCAL code in the appropriate section, and it should be included when any call is made initially to a receiving station.

VHF R/T channels

To prevent radio interference between two or more ground stations, a geographical separation is determined between international services using the same or adjacent frequencies. Limits to the height and range at which aircraft should use these frequencies are specifically laid down, and except for emergencies or unless otherwise instructed by ATC, pilots should adhere to these limits. For international aerodrome services, en route sector limits for use of these frequencies are:

TOWER	25 nm	4,000 ft
APPROACH	25 nm	10,000 ft

Other aerodrome services frequency utilisation is greater, and to satisfy the greater demand for frequencies many neighbouring aerodromes may use the same or adjacent frequencies for similar services. To prevent interference, use of these frequencies by aircraft is limited by height and distance restrictions from the aerodrome. Communication on TOWER, AFIS and Air/Ground facilities are limited to 10 nm and 3,000 ft maximum and, as far as possible, to a height of 1,000 ft in the immediate vicinity of the aerodrome.

The Aeronautical Fixed Service

The Aeronautical Fixed Service is a telephone service which includes the Aeronautical Fixed Telecommunications Network (AFTN) for supporting operational services connected with air traffic operations for the safe operation of air navigation and the regular, efficient and economical operation of air services.

The Aeronautical Broadcast Service

Meteorological, navigation and aerodrome information is provided by broadcasts on the Aeronautical Broadcast Service. The broadcast details are listed under the individual aerodrome or ATC unit in the COM 2 and MET 3-1 sections of the *UK Air Pilot*.

General

Portable telephone interference

The use of portable telephones is not permitted in aircraft because they cause interference to radio and navigational equipment. It also contravenes both the aircraft and the telephone user's licence conditions.

High-power transmitter interference

High-power transmitters and high-intensity radio transmissions can cause interference and possibly damage to aircraft electronic equipment. This may cause the navigation information to be unreliable. If, beyond the vicinity of the ground transmitter, interference is troublesome or experienced, pilots are requested to file a Ground Fault Report Form CA 647 to include the following:

a Frequency on which interference occurred
b Position and height of aircraft
c Aircraft registration letters
d Date and time of interference
e Description of interference, (e.g. speech, language, music, etc.).

Aircraft radio operation

Except in the cases of gliders and student pilots under instruction, the aircraft radio installation must be operated by a person who is licensed to do so. Except for telephonic stations, a telegraphic station is required to keep a radio log book for all transmissions. When SELCAL is authorised for use by ATC, a listening watch does not need to be maintained in controlled airspace; otherwise it is mandatory at all times. The licence permits messages to be transmitted and received for:

a Emergency purposes
b Details relating to the aircraft's flight
c Public correspondence authorised under that licence.

When flying public transport aircraft in controlled airspace, pilots and flight engineers must not use hand-held microphones below FL150, or when taking off and landing.

When acknowledging ATC messages, pilots must read back in full:

a ATC clearances
b Heading and level instructions
c SSR operating instructions
d Altimeter settings

e VDF information
f Frequency changes.

When ground navigational aids are being used, an identification signal is sent out. If the identification signal is 'TST', this must not be used for operational purposes, as it is only a test transmission.

Chapter 10
Meteorology

The information in this chapter covers the relevant items particular to the Air Law requirements, and in no way covers the meteorological syllabus for examination.

The Meteorological Organisation

There are four types of meteorological offices throughout the UK, and they are organised in the following manner:

1 ***Main offices*** These are located at major civil aerodromes and ATCCs, and they are open 24 hours. They supply forecasts and other met information to their own aerodromes and to their attached observing offices, while controlling and advising their observing and subsidiary offices. Charts are normally prepared every three hours, and as necessary during the intermediate period.

2 ***Subsidiary offices*** These are located at civil aerodromes of intermediate importance, and their hours of operation are subject to local requirements. The issue of forecasts and other met information is carried out in a manner similar to the main offices.

3 ***Observing offices*** Minor civil aerodromes are the locations for these offices, which issue weather reports but not forecasts. Forecasts and other met information can be obtained from the main and subsidiary offices to which they are attached. They are generally open to meet the local requirements, but are open for 24 hours when classed as synoptic reporting stations.

4 ***Meteorological Watch Offices*** (MWOs) The UK provides a meteorological watch over its designated airspace, and an MWO is based at the main office of the ATCC. These areas within the UK are listed as follows with their relevant MWOs:

- **a** Scottish FIR/UIR is covered by the Glasgow Weather Centre
- **b** London FIR/UIR is covered by Bracknell CFO
- **c** Shanwick FIR/UIR is covered by Bracknell CFO.

Apart from liaising with other met offices in the area, the functions of the MWOs are to advise ATC on diversions, provide en route forecast services and initiate SIGMET messages.

Observing Routine

Meteorological observations are made every half hour at principal aerodromes and hourly at other aerodromes. Any sudden deteriorations and improvements are the subject of special reports.

Types of Service Provided

Pre-flight briefing

The principal method of briefing aircrew, in the UK, of meteorological conditions, is by self-briefing. This is accomplished by using facilities, information and documentation available or displayed in aerodrome briefing areas as a matter of routine. An AIRMET telephone recording service provides briefing for pilots, available for flights below 15,000 ft AMSL. This principal method of briefing requires no prior notification.

If the pilot considers that the self-briefing is inadequate for the intended flight, or it is unavailable, a Special Forecast may be provided.

Special Forecasts and specialised information

When required, Special Forecasts may be requested from the Forecast Office to cover a specific period for a designated route or part of that route. When Special Forecasts are required from the Forecast Offices, prior notification is generally required as follows:

- **a** At least 2 hours before the time of collection for flights up to, and including 500 nm
- **b** At least 4 hours before time of collection for flights of over 500 nm.

Special Forecasts are normally issued to the aerodrome of departure by AFTN, Telex, or DOCFAX. However, pilots may telephone the Forecast Office for the forecast when the normal method is unavailable.

Emergencies, in-flight forecasts and properly notified forecast requirements, are given priority by the Forecast Office. Other requests at busy periods could be delayed, and it is therefore in the interest of all concerned that the maximum possible notice is given.

Prior notification is not normally required by the Forecast Office for take-off forecast information containing surface wind, temperature and pressure.

Organisers of specialised aviation events requiring meteorological information may find that it is not covered by the AIRMET service or given as Special Forecasts. In the case of the following specialised aviation uses, provision for meteorological information can be provided on prior request.

1 Gliding, hang gliding, balloon and microlight organisations can obtain low-level wind, lee-wave, QNH, temperature and thermal activity forecasts. (At least 2 hours advance notice is required.)

2 Offshore helicopter operators can be provided with forecast winds and temperatures at 1,000 ft AMSL, plus information on sea state and temperature and on possible airframe icing conditions.

3 Special aviation events can be given meteorological information when routine forecasts are inadequate.

NOTE: Approval for items **2** and **3** above should be made to the CAA at least six weeks in advance, stating meteorological requirements.

Meteorological charts

Meteorological charts are normally transmitted through a facsimile network known as CAMFAX to all major aerodromes. They cover:

a Low- and medium-level flights within the UK and near Europe
b Medium- and high-level flights to Europe and the Mediterranean
c High-level flights to North America and the Middle and Far East.

Charts for other areas that are not routinely available can be obtained by prior request from Bracknell CFO. Geographical and vertical coverage of charts, their availability, their time of issue and validity are all laid out in the Met section of the AIPs. Some aerodromes have another facsimile method, known as DOCFAX, to supply chart and met information.

Broadcast text met information

Various met broadcasts are made by teleprinter throughout the UK and internationally in text form. These are:

a METAR	Aerodrome meteorological reports
b TAF	Aerodrome Forecasts
c SIGMET	Warnings to flight safety and volcanic activity reports.

Within the UK this information is distributed through the following communication channels.

a OPMET
b AFTN
c Autotelex.

Information regarding these communication channels and networks are laid down in the MET section of the AIPs. When scattered Selected Special Reports are made beyond the aerodrome of origin they are known as SPECI reports.

AIRMET service

The AIRMET service covers three areas of routine forecasts, in plain language, which cover the UK and near European Continent. These three areas are covered as follows, and are illustrated in Fig. 10-1 on page 116:

a Scottish Region
b Northern Region
c Southern Region (covering the Channel Islands and parts of France).

The vertical coverage is from the surface up to 15,000 ft AMSL, with winds and temperatures up to 18,000 ft AMSL. This is provided through the public telephone system in spoken form, or in printed form through the Aeronautical Fixed Telecommunication Network (AFTN) and telex systems. All three of the AIRMET forecasts, as well as selected Aerodrome Forecasts, are provided on a 24-hour basis through PRESTEL (British Telecom videotex service) for aviation users by the Meteorological Office. To obtain further information of the coverage of this service, contact should be made with the Met Office or through PRESTEL key *20971#.

Forecasts are issued four times daily at six-hourly intervals, and are amended as necessary. Forecasts will reflect the contents which are current at the time of issue, of SIGMETS, or of any amendments to the forecasts.

In-flight procedures

MWO procedures, which the En Route Forecast Service supplements when necessary, supply information to aircraft in flight as required. Information can also be supplied from meteorological broadcasts or the appropriate ATS units when required,

Within its own and adjacent FIRs an MWO is responsible for the preparation and dispersal of SIGMETs to the appropriate ACC/FICs within its area. Each ACC/FIC should then warn all aircraft in flight within its area of any of the phenomena covered under the SIGMET for routes ahead, up to 500 nm or 2 hours' flying time.

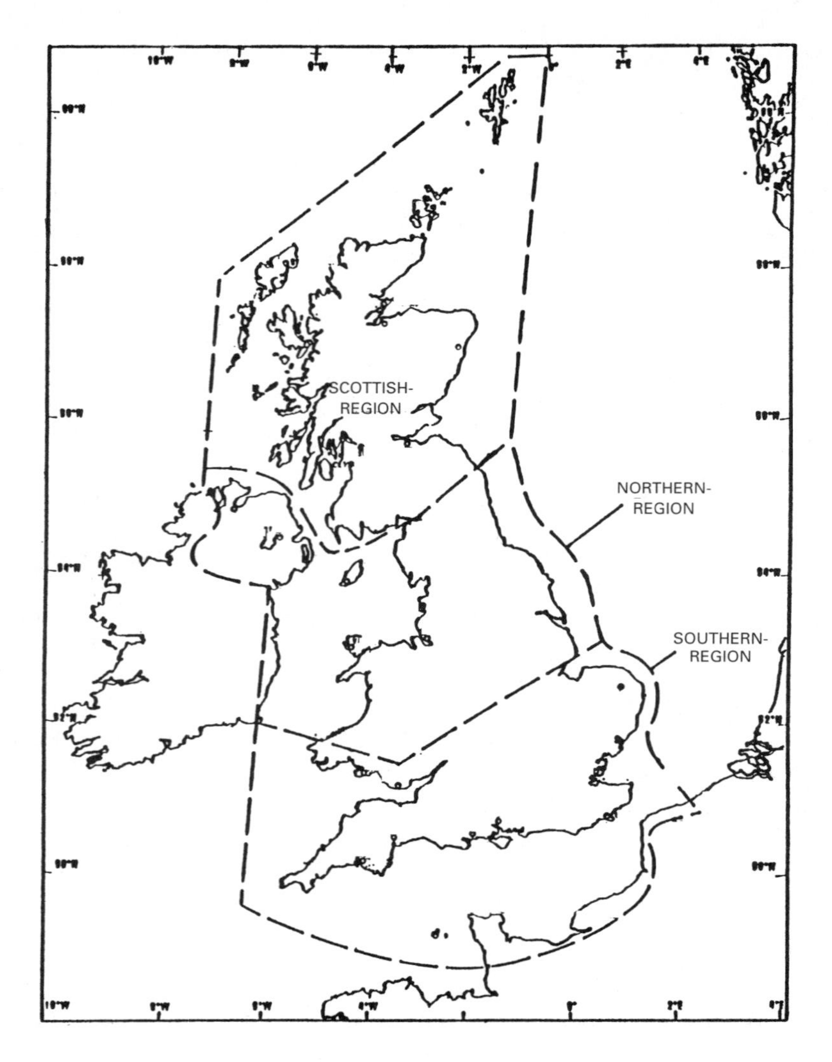

Fig 10–1 AIRMET Boundaries

SIGMET PHENOMENA

1 At subsonic cruising levels (SIGMET):

a Active thunderstorms (as defined in the AIPs)
b Tropical cyclones
c Severe line squalls (surface squalls of 50 kt or more, accompanied by other SIGMET phenomena)
d Heavy hail
e Severe turbulence
f Severe icing
g Marked mountain waves (vertical air currents above 500 ft/min)
h Widespread sandstorm/dust storm
i Volcanic ash cloud.

2 At transonic and supersonic cruising levels (SIGMET SST):

a Moderate or severe turbulence
b Cumulonimbus clouds
c Hail
d Volcanic ash clouds.

NOTE: The transonic and supersonic cruising levels mentioned above refer to FL250–FL600 for the London and Scottish UIRs, and FL400–FL600 for the Shanwick OCA.

The validity period of SIGMETS is not normally more than 4 hours, except for volcanic ash cloud, which may be extended up to 12 hours' validity. An in-flight En Route Forecast Service is available by prior arrangement and only in exceptional circumstances.

Aerodrome weather information

An aircraft may obtain aerodrome weather information by one or more of the following methods:

a VOLMET broadcasts, which contain:

i	Surface wind	**v**	Cloud
ii	Visibility	**vi**	Temperature
iii	RVR	**vii**	Dew point
iv	Weather	**viii**	QNH

b ATIS broadcasts
c By request to an ATS unit (but, whenever possible, only if the required information is not available from a broadcast).

Diversion

If an aircraft has cause to divert to an aerodrome along a particular route where no forecast has been provided, the pilot may request forecast information from the appropriate ATS unit. The associated Forecast Office will then provide the necessary forecasts to facilitate the best diversion choice available.

SNOCLO

The word 'SNOCLO' will be added to the end of aerodrome reports (VOLMET) to indicate that the aerodrome is temporarily closed because the runways are unusable owing to heavy snow falls, or that snow clearance is being undertaken, thereby making landings and take-offs impossible.

Marked temperature inversion

A warning of marked temperature inversion is issued at certain aerodromes whenever a temperature difference of 10 degrees C or more exists between the surface and any point up to 1,000 ft above the aerodrome. At aerodromes so equipped, this warning is broadcast on the arrival and departure ATIS, and at aerodromes where ATIS is unavailable it is passed by radio to aircraft before take-off and to arriving aircraft as part of the aerodrome meteorological report.

Flight over and in the vicinity of High Ground

The expression 'HIGH GROUND' is used in meteorology to describe mountains, hills, ridges, etc., which rise to heights in excess of 500 ft above nearby low-lying terrain.

Aerodrome warnings

Aerodrome warnings are issued when one or more of the following phenomena occurs, or is expected to occur:

- **a** Gales (when the surface wind is expected to exceed 33 kt, or if gusts are expected to exceed 42 kt) or strong winds according to locally agreed criteria
- **b** Squalls, hail or thunderstorms
- **c** Snow warnings, which include expected:
  - **1** Time of beginning, duration and intensity of fall
  - **2** Depth of accumulated snow
  - **3** Time of thaw.
- **d** Frost warnings, if any of the following conditions are expected to occur:
  - **1** A ground frost with air temperatures not below freezing point
  - **2** The air temperature above the surface is below freezing point (air frost)

3 Freezing precipitation
4 Hoar frost, rime or glaze deposited on parked aircraft.

e Fog (normally when visibility is expected to fall below 600 m).

Aerodrome warnings are normally notified by AFTN, Telex or telephone message to the aerodrome.

CAVOK

The term CAVOK replaces the RVR, visibility, cloud and weather group when the following conditions exist:

a 10 km or more visibility;
b No cloud below 5,000 ft or below the highest Minimum Sector Altitude, whichever is greater, and no cumulonimbus;
c No precipitation, thunderstorm, shallow fog or low drifting snow.

QNE

When atmospheric pressures are so exceptionally low that it is impossible to set QFE or QNH on the aircraft's altimeter, a runway QNE may be requested. The QNE is when the altimeter, set to 1013.2 mb, indicates the altitude with the aircraft on the aerodrome or runway. In this circumstance the pilot should take extra care regarding terrain and obstacle clearance, and also vertical separation from other aircraft using QFE or QNH.

Runway visual range

The visual range along a runway is known as the RVR, and an assessment can be made either by a human observer or by an instrument, when it is known as an IRVR. The IRVR is more accurate than a visual assessment by a human observer.

Whenever the horizontal visibility of a runway is reported to be less than 1,500 m an RVR assessment is made. With the IRVR system this assessment is made when the visibility is reported as 1,500 m or less, the difference being due to the accuracy of the instrument system compared with the human eye.

RVR is determined by markers in daylight, and by lights at night or in dull, foggy or smoky conditions. To assess runway visibility under these conditions, the observer is positioned 76 m from the runway centreline and counts the lights or markers visible from that point. Markers are placed at intervals of 50 m for the first 500 m, and at 100 m intervals from 500 m up to 1,500 m. The markers/lights are positioned 9 m to the left of the runway.

The IRVR system may be of the three-site or two-site type, depending upon the aerodrome. In the three-site system, if a single transmissometer (see Fig. 10-2 on page 120) fails, the system is still serviceable, and the remaining site information is passed on to the pilot. If two transmissometers fail, the remaining value will be passed on to the pilot provided it is not the stop-end value, in which case the system is classed as unserviceable. A two-site system gives the touchdown zone and the stop-end values. If the touchdown value fails, the system is considered unserviceable for that runway direction.

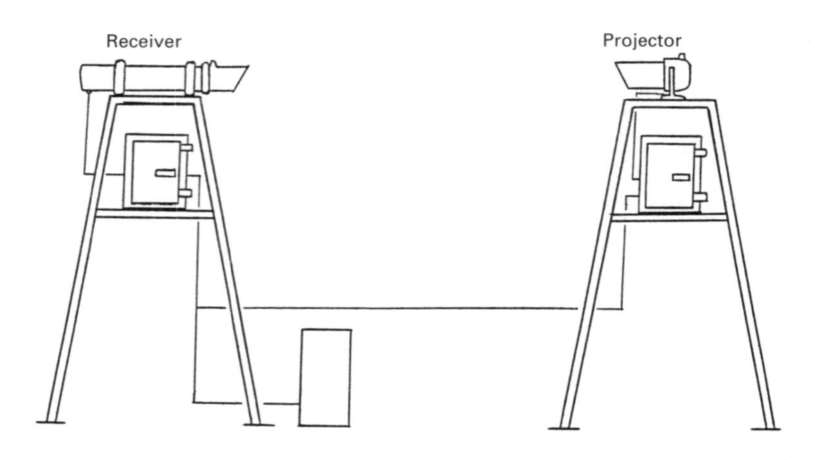

Fig 10–2 Transmissometer

Reporting of the RVR and IRVR within the UK is accomplished on an incremental scale of 25 m between 0 and 200 m, 50 m between 200 and 800 m, and 100 m above 800 m up to the maximum. Every incremental scale is not always able to be reported, as some IRVR systems and most human observer systems are unable to comply.

Air-to-Ground Reporting

Routine aircraft observations

In-flight weather observations, for air-to-ground transmissions, are not required in the London or Scottish FIR/UIR or the Shanwick FIR. Certain routine meteorological observations are required within the Shanwick OCA (Atlantic crossings). These are normally transmitted at the same time as position reports, and in some instances at intermediate midpoint reports. These meteorological reports are generally sent in AIREP code.

Special aircraft observations

Special meteorological reports are sent, using the prefix AIREP SPECIAL, as a requirement in any UK FIR/UIR/OCA whenever one of the following occurs:

a Severe turbulence or severe icing is encountered
b Moderate turbulence, hail or cumulonimbus clouds are encountered during transonic or supersonic flight
c Meteorological conditions are encountered which may affect the safety or the efficiency of other aircraft operations. These conditions could be any of those listed under SIGMET, or any adverse conditions during climb-out or approach not previously forecast or reported
d Special requested reports by the meteorological service covering that flight
e Special agreement made between the Meteorological Authority and the aircraft operator.

CHAPTER 11
SEARCH AND RESCUE

Introduction

As a member state of ICAO, the United Kingdom is committed to providing coverage for search and rescue (SAR) in the event of an accident, malfunction or other problem that causes an aircraft either to force/crash land or 'ditch' at sea. This must be on a 24-hour basis, and covers the UK overland areas and the adjacent sea area between 030 degrees W over the North Atlantic (except for the Shannon FIR), to the midpoint section between the UK and the European mainland. This area is illustrated in Fig. 11-1. The UK area of responsibility is divided into two sections, the northern sector being covered from the Rescue Co-ordination Centre (RCC) at Edinburgh, and the southern area being covered by the RCC at Plymouth.

The SAR service is a combined military and civil operation, the RCCs having military operational control, while the SAR helicopters contracted to the Department of Transport (DoT) are under the operational control of HM Coastguard.

The rescue organisation

The rescue organisation is a combination of the ATCCs and RCCs. These units work in close co-operation with each other to allow the most efficient usage of manpower and aircraft. Any area, either land or sea, must be free from excessive aerial activity. This is taken care of by the ATCC, which allows the RCC to complete its task unhindered and in a correct and safe manner. To achieve efficient operation of a SAR mission, various types of services are employed, and these are divided into military and civil elements.

The military element consists of:

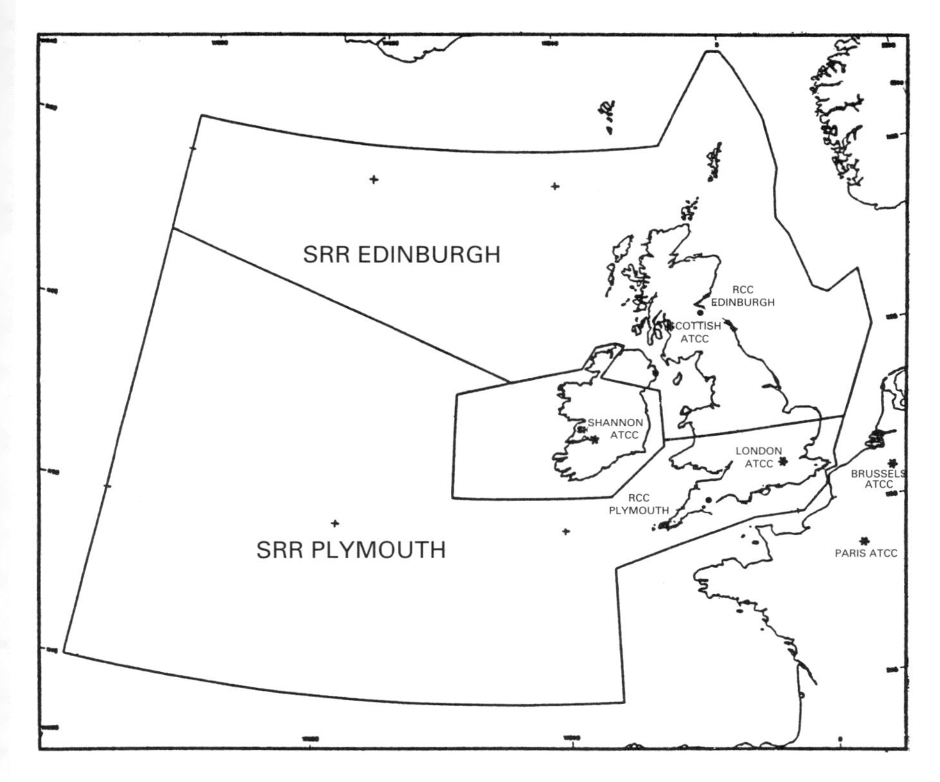

Fig 11–1 UK Search & Rescue Regions

a RAF and RN fixed-wing aircraft and helicopters
b RAF Mountain Rescue Teams (MRT)
c RN ships and helicopters non-SAR-dedicated and RAF aircraft non-SAR-dedicated
d Army, RN and RAF personnel
e Neighbouring RCC.

The civil element consists of:

a DoT helicopters
b HM Coastguard
c Royal National Lifeboat Institution
d Civil police, fire and ambulance services
e Civil aircraft

f Merchant vessels
g Civilian mountain rescue teams
h British Telecom International Coast Radio
i Cosmos Rescue System (COSPAS) and the Search and Rescue Satellite Aided Tracking System (SARSAT), both satellite distress-alerting-systems.

The military specialised landplanes carry Lindholme droppable survival aids which include radio, water, food and multi-seat inflatable liferafts. These aircraft are normally RAF long-range maritime patrol (LRMP) aircraft, but RAF transport aircraft can also be used for SAR. All of the SAR helicopters are specially equipped with winching gear.

The mode of operation for maritime SAR is for the RAF LRMPs to locate and initially equip personnel in distress in the water, and further to direct shipping and helicopters to effect the rescue.

Distress frequency allocations

Four basic frequencies are used for the distress calling. These are 121.5 mHz, 243.0 mHz, 2182 kHz and 500 kHz. The 500 kHz is a carrier-wave signal used for homing purposes by RAF LRMP aircraft, whereas the other three frequencies have speech facility.

The scene of search uses various frequencies as listed in the *UK Air Pilot*. However, the main civil frequency is 123.1 mHz, and the NATO frequency is 282.8 mHz.

A more modern distress frequency is 406.025 mHz, which is transmitted through the COSPAS/SARSAT satellite units. This method has coded facilities which enable the RCCs to pinpoint the signal within minutes of it being sent, to a far greater accuracy of position pinpointing.

Alert phasing

The sequence of alerts for overdue aircraft covers three basic phases. These are:

a Uncertainty phase
b Alert phase
c Distress phase.

Although this phase sequence is basically for overdue aircraft, if a known alert condition exists such as a MAYDAY call, then the ATCC may go straight to the ALERT or DISTRESS phase, as the case may be.

Uncertainty phase

This period is initiated by the ATCC:

a If no communication is received within 30 min after a communication contact should have occurred
b If no reply is heard to a given transmission
c If an aircraft fails to arrive within 30 min of the last notified ETA or that estimated by the ATS, whichever is later.

At this stage the RCC is informed and the necessary rescue units are placed on alert while information is collected and evaluated.

Alert phase

This phase will be initiated immediately when the information received indicates that the operational efficiency of the aircraft has been compromised. It will also be immediately initiated when an aircraft has been given clearance to land and fails to do so within 5 min without any obvious reason for doing so (*i.e.* overshoot, etc).

The uncertainty phase will be upgraded to an ALERT PHASE when there is no further news of the aircraft and/or subsequent attempts to establish communication fails.

Distress phase

This phase will be initiated immediately when an aircraft is known to have been forced to land or ditch in the sea. The ALERT PHASE will be upgraded to the DISTRESS PHASE when further attempts fail to establish contact, or when the known fuel reserves are calculated to be exhausted. The accuracy of the endurance stated on the flight plan plays an important part in this calculation.

During the ALERT PHASE the state of readiness may be shortened from the normal readiness state while awaiting the DISTRESS PHASE to be reached. The normal states of readiness for SAR units are as follows:

SAR landplanes	60 min (day or night)
SAR helicopters	15 min (day)
Sea King/DoT S-61N	45 min (night)
RAF Wessex	60 min (night)
MRTs	60 min (day or night)

It is in the interests of all aviators to help themselves by preparation and assistance. To begin with, if an aircraft has no radio and a flight is intended to exceed a distance of 10 nm from the coast, or if an intended flight is over difficult, sparsely populated or mountainous areas, a flight plan should be filed. This is intended to let everyone know that the aircraft is airborne and is intending to land safely at a known destination. If the intended destination is not on the AFTN network, it is in the pilot's interest to notify a responsible person at the destination airfield of the ETA.

Pilots flying over the above terrain should ensure that the correct and sufficient survival equipment is carried, which should include a survival radio.

Within the UK there are many areas where SAR is difficult and a survival situation can occur. Every year several people die owing to becoming lost, not having sufficient survival equipment, or just by being totally unprepared for the conditions in the following areas:

the Scottish Highlands;
the Hebrides, Orkneys and Shetlands;
the Pennine range;
the Lake District;
the Yorkshire Moors;
the Welsh mountains;
the Peak District of Derbyshire;
Exmoor and Dartmoor.

Emergency calls and procedures

Emergency procedures

During a flight, emergencies can occur at any time and without warning. It is important that pilots have a thorough knowledge of emergency procedures, so that they can react swiftly and correctly if an emergency occurs. The emergency procedures that apply to emergency calls and messages, emergencies involving other aircraft, and emergencies with unserviceable communication equipment, are now considered.

Degrees of emergency

Emergencies are classified into two degrees, depending upon their seriousness. The degrees of emergency are:

DISTRESS A distress emergency occurs when an aircraft is threatened by serious danger and needs immediate assistance. Examples of a distress emergency are:

- An aircraft ditching into water
- An aircraft crash landing
- The flight crew abandoning an aircraft.

URGENCY An urgency emergency occurs when a situation endangers the safety of an aircraft, a person on board an aircraft, or another aircraft. Examples of an urgency are:

- An aircraft is lost
- An aircraft has a low fuel condition
- An aircraft is experiencing partial engine failure.

Distress and urgency radio transmissions

An emergency radio transmission used with distress and urgency emergencies consists of two parts:

Emergency call
Emergency message

DISTRESS EMERGENCY CALL When an aircraft requires immediate assistance because the safety of the aircraft and its occupants is in jeopardy, a distress emergency call is made by transmitting the pro-sign MAYDAY three (3) times, followed by the aircraft callsign. An example of a distress call for aircraft XXX 123 is:
MAYDAY, MAYDAY, MAYDAY, XXX 123
or the emergency can be signified by:

1 Firing a red parachute flare, or
2 Red pyrotechnic lights fired at short intervals, or
3 An SOS signal in Morse (where possible by light signals or sound).

When an aircraft is signalling to surface craft, a visual signal is more effective than a sound signal.

URGENCY EMERGENCY CALL The emergency call for an urgency emergency is made by transmitting the pro-sign PAN three (3) times, and then the aircraft callsign. An example of an urgency call is:
PAN, PAN, PAN, XXX 123
EMERGENCY MESSAGE
The emergency message is transmitted after the emergency call, and consists of the following (as time permits):

- Name and station addressed (where appropriate)
- Callsign and type of aircraft
- Nature of emergency
- Intention of aeroplane captain
- Present position, flight level/altitude and heading.

In the UK only, and when possible:

- Pilot's qualifications, and instrument ratings if applicable.

An aircraft with a radio may call a 'PAN PAN' or 'MAYDAY' call, depending on the nature of the emergency and seriousness of the situation. The call should be made on the frequency in use, and although directed at the ATCU with whom contact is already established, it should be an open call to all persons. If the pilot is unable to make contact on the frequency in use, a call should be made on 121.5 mHz, where an aid service is continuously available. In either case, where a transponder is fitted, a code of 7700 (emergency) should be selected.

If a 'MAYDAY' call is made and the situation has improved sufficiently, the emergency call can be downgraded to a 'PAN' call, or it can be upgraded from a 'PAN' call to a 'MAYDAY' if the situation deteriorates. Once the emergency situation no longer exists, the controlling authority should be informed so that the SAR service can be secured and resume its normal state of readiness. The message must be transmitted on the same frequency or frequencies on which the original message was transmitted.

Assistance required

When an aircraft is flying over sparsely populated areas or over water and the operational efficiency of the aircraft has been compromised, assistance may be requested by the aircraft commander. An escort can be provided by contacting the ATCC, who will make the arrangements. Owing to the limited number of SAR aircraft, a request for an escort should only be made if essentially required. This escort service is provided free of charge for legitimate requests.

Observing another aircraft with an emergency

An aircraft observing another with an emergency is to take the following actions, if possible

- Keep the aircraft in sight and switch your SSR to emergency
- If no distress transmission is heard, transmit a distress message containing all information
- Comply with instructions from ATC or controlling authority on the ground and maintain visual contact with the aircraft.

Intercepting a distress call

An aircraft that intercepts a distress call is, if possible, to:

- Take a bearing on the transmission and plot the aircraft's position
- Monitor the frequency used by the aircraft in distress
- Proceed to the aircraft's position, observe for visual signals, and follow ground authority instructions (for as long as possible).

Unserviceable communications equipment

In the event that you become lost or uncertain of position (an urgency emergency) and are unable to communicate with ATC, there are standard procedures to follow. These procedures can notify ATC that a problem exists and that you have:

- An unserviceable transmitter, or
- An unserviceable transmitter and receiver.

Distress procedure for radar identification

If a pilot is able to fly a pattern which could be identified by a ground radar station, the following action should be taken.

Unserviceable transmitter (Fig. 11-2)

If your transmitter is unserviceable:

- Switch the SSR to the emergency code (7600)

Continue to transmit on the initial frequency and monitor the emergency frequency.

- Switch radio to emergency frequency and fly a RIGHT triangular pattern of:
  One-minute legs (airspeed above 300 kt), or
  Two-minute legs (airspeed below 300 kt)

- Complete TWO RIGHT triangular patterns frequently.

ATC radar will observe these patterns and attempt to transmit instruction on the emergency frequency. Therefore listen for instructions from a ground radar station.

Unserviceable transmitter and receiver (Fig. 11-2)
If the transmitter and receiver are both unserviceable:

- Switch the SSR to the emergency code (7600), and continue to monitor the emergency frequency
- Fly a LEFT triangular pattern of:
  One-minute legs (airspeed above 300 kt)
  Two-minute legs (airspeed below 300 kt)
- Complete TWO LEFT triangular patterns and repeat the patterns frequently.

ATC will observe the patterns and provide an aeroplane to fly to your position and assist you. Therefore watch for interception by a shepherd aircraft.

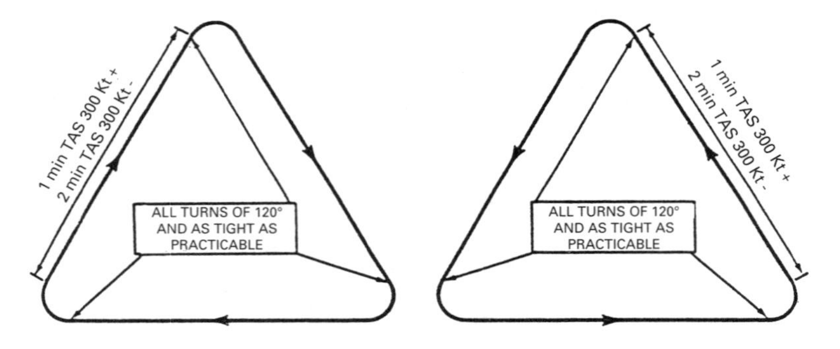

Fig 11–2 Unserviceable Communication Equipment Procedure

NOTE: **1** Switch IFF/SSR to Mode 3 Code 76/7600
2 Fly two patterns, resume course, repeat at 20 min intervals
3 Guard emergency frequencies.

En route communications failure procedures

When flying under a filed flight plan, apart from carrying out the identification procedures, the following procedures are to be carried out:

1 If in VISUAL METEOROLOGICAL CONDITIONS (VMC)
a Continue to fly in VMC
b Land at the nearest suitable aerodrome
c Report the aircraft's arrival position and situation to the appropriate ATC unit.

NOTE: 'Suitable aerodrome' means the nearest aerodrome suitable for the operation of the aircraft and at which radio telephonic or telephonic communication facilities are available for reporting the aircraft's position.

2 If in INSTRUMENT METEOROLOGICAL CONDITIONS (IMC)
If in IMC, or when the weather conditions are such that is does not appear possible to complete the flight under the VMC conditions, the aircraft shall:
a Proceed according to the current filed flight plan, flying to the most suitable indicated navigational aid operating at the aerodrome at which the landing was intended. Hold over this navigational aid until required to begin descent
b Start descent from the navigational aid, from where the holding fix is operating, at the expected approach time (EAT) last received and acknowledged, or as close as possible to this time. If no expected approach time has been received and acknowledged, the descent should begin at, or as close as possible to, the ETA as shown on the filed flight plan and revised in accordance with the current flight plan
c Complete a normal instrument approach procedure as stated for the indicated navigational aid
d Land, if possible within 30 min after the ETA given, or the last acknowledged expected EAT, whichever is later.

NOTE 1: During these procedures it is the pilot's responsibility to keep a sharp lookout for other aircraft in the pattern, the hold, the descent and the circuit for landing.
NOTE 2: When the aircraft is part of the aerodrome traffic, the pilot should keep a watch for the instructions which may be given by visual signals.
NOTE 3: The landing runway and direction can be obtained by observing the landing 'T' or by observing the other aircraft in the traffic pattern.

Controlled descent with unserviceable microphone

If an aircraft is above cloud with an unserviceable microphone, but can receive a ground station, a normal controlled descent can be carried out using the following procedure:

Procedure by pilot

If in an emergency, select the appropriate distress frequency and press the transmitter button FOUR times. This will appear on the CR D/F screen as four short transmissions representing the Morse code for the letter 'H', for HOMING.

During all subsequent procedure, strict use is to be made of the following code:

1 One click means 'YES' (or acknowledgement)
2 Two clicks mean 'NO'
3 Three clicks mean 'SAY AGAIN'
4 The letter 'X' in Morse code (—..—)means 'an additional or greater degree of emergency'.

ATC will pass instructions and will detail the acknowledgement to be given (e.g. GIVE THREE SHORT SHARP TRANSMISSIONS COMMENCING THE INBOUND TURN). Unless otherwise instructed by ATC, the following stages in the controlled descent pattern are to be indicated by a two-second transmission:

1 Overhead turn complete
2 Steady on inbound heading
3 Steady at 'CHECK HEIGHT'
4 Steady at 'BREAK-OFF HEIGHT'
5 Airfield in sight.

Using the foregoing codes, the controller can interrogate the pilot by asking questions which permit 'YES' or 'NO' answers. The aircraft can be homed overhead and a normal descent carried out. Pilots hearing the 'SPEECHLESS AIRCRAFT' are to maintain R/T silence on the frequency being used as far as possible.

Types and use of survival equipment

There are many types of survival equipment. These would require a separate book, and therefore a brief write-up on the basic items of survival equipment and signals are given here. Survival equipment and its uses can be separated into two categories. These are:

1 The crash-landing situation.
2 The ditching situation.

Let us first look at the crash-landing situation, in which environmental features may assist in survival, and in the location and rescue of the survivor.

To begin with, all aircraft on every flight should, for safety reasons, carry the minimum equipment as follows:

a Distress flares and/or cartridges
b Emergency radio/survival beacon/aircraft radio
c Heliograph or mirror
d Ground/air emergency code
e Emergency torch
f Emergency ration pack
g First aid kit.

A dinghy survival pack is also a helpful item of equipment in a crash/forced landing. The environment may help in providing shelter, water and food. It can also be used in making the ground/air signals or a triangle of fires. The fires can be made to produce flame for location at night, or smoke for location by day. Both are recognised international distress signals.

Ditching restricts the number of location aids available, but if used in the correct manner they will help the survivor to be rescued. The main location aids for a survivor at sea are:

a Lifejacket
b Dinghy
c Lights
d Emergency radio/beacon
e Rocket/parachute flares
f Miniflares
g Heliograph
h Fluorescent dye marker
j Whistle.

Another recognised international distress signal is flying a flag with a ball above or below it. This method of signalling is very restricted, and in some cases impossible for the ditched aircraft victim. It is more commonly used for a boat in distress, or where a mast can be erected.

Duration and use of radio

Most emergency radio/beacons have a battery with a short-term life of approximately 24 or 36 hours, depending on the model. Some survival situations may last longer than this, so economic use of the radio is essential. A call should be put out initially, of approximately 2 min duration. The call should be repeated at approximately 30 min intervals for the first two hours, and at two-hour intervals thereafter. The exceptions to this are when the survivor is on a well used air route, or when an aircraft is heard in the vicinity of the survivor.

Some old-style dinghy radios which operate on 500 kHz send out a Morse signal which can be picked up by LRMP aircraft. The power for this type of radio is generated by turning the crank handle. The 'watch' period for this frequency is every 30 min from 15 min past the hour, and 45 min past the hour, and lasts for approximately 3 min. Similarly, transmissions on 2182 kHz should be made during the international silence periods of three minutes starting at H and H +30.

RAF search techniques

When RAF aircraft are searching for survivors at night, a certain pattern is adopted. This pattern is:

a A single green pyrotechnic is fired from the search aircraft at approximately 5 to 10 min intervals
b After 15 sec from the survivor seeing the signal, the survivor should fire a red pyrotechnic, followed by a second after a short period
c If the search aircraft appears to be getting off track, an additional pyrotechnic should be fired
d An additional pyrotechnic should be fired when the search aircraft is approaching overhead.

During the day or night the survivor may be asleep or semi-conscious when an aircraft is heard in the vicinity, and a missed opportunity might occur in sending up flares. If it is a search aircraft, which should be at approximately 1,000–2,000 ft AMSL, it will be following a special search pattern. This search pattern will mean that at some stage it will pass the area of the survivor again. The main search patterns are as follows:

a Clover-leaf pattern
b Expanding square search
c Creeping line ahead.

GROUND – AIR VISUAL SIGNAL CODE FOR USE BY SURVIVORS

No.	*Message*	*Code Symbol*
1	Require assistance	V
2	Require medical assistance	X
3	No or negative	N
4	Yes or affirmative	Y
5	Proceeding in this direction	↑

GROUND – AIR VISUAL SIGNAL CODE FOR USE BY RESCUE UNITS

No.	*Message*	*Code Symbol*
1	Operation completed	LLL
2	We have found all personnel	LL
3	We have found only some personnel	++
4	We are not able to continue. Returning to base	XX
5	Have divided into two groups. Each proceeding in direction indicated	⇄
6	Information received that aircraft is in that direction	→ →
7	Nothing found. Will continue to search	NN

Chapter 12
Facilitation

Introduction

Aircraft on international flights to and from countries outside Great Britain may cross the coastline or land frontiers at any point, but they are subject to the rules and regulations in current operation which include those which are contained in the UK AIP, and are found in the RAC section. These rules and regulations cover Immigration and Customs as well as the individual flying restrictions local to each airport or aerodrome. This chapter covers the main points in reference to these rules and regulations, and is purely intended as a guide to understanding them.

Arrivals

Unless special permission has been granted before landing within the UK, an aircraft entering the UK from abroad must land at a designated Customs and Excise airport/aerodrome or airfield that is listed in the UK AIP, FAL section. This requirement is to be carried out every time an aircraft enters the UK:

- **a** For the first time after its arrival in the UK; or
- **b** At any time while it is carrying passengers or goods brought in that aircraft from a place outside the UK and not yet cleared.

When requests for the referred special permission are made, it should be made by or on behalf of the aircraft commander through the aerodrome operator to the local Customs and Excise office.

Aircraft – liability for Duty

Aircraft from abroad arriving in the UK are all subject to Customs Duty. However, the following are exempt from this duty:

a UK based aircraft which have undergone only routine servicing repairs and not major repair or any renovation whilst abroad
b Aircraft on international scheduled services which are normally based outside the UK
c Aircraft temporarily imported into the UK which are operated by resident persons or companies based abroad. Certain conditions affect the duty relief to which these are subject, and these conditions are as follows:

1 The aircraft must not be lent, sold, or otherwise disposed of while in the UK.
2 The aircraft must not be used for demonstration purposes with a view to sale of similar aircraft.
3 The aircraft must not be used for the purposes of remuneration in the carriage of persons or goods between places within the UK.
4 Some aircraft being imported for repair or overhaul may qualify for relief, but those imported for aerial work (i.e. crop-spraying) do not.

Departures

An aircraft may not depart from the UK to any place or area outside the UK except from a Customs and Excise airport unless special permission has been obtained from the Commissioners of Customs and Excise. Requests for such permission should be made through the aerodrome operator by or on behalf of the aircraft commander.

When special permission has been granted, the requirement of the ANO for aircraft engaged in the public transport of passengers to use a licensed aerodrome must still be adhered to.

After a departing aircraft has been cleared it must not land again within the UK at any place other than a Customs and Excise airport stated in the General Declaration. The only exception to this is when an aircraft is forced to land.

The owners of UK registered aircraft should present to the Customs and Excise officer a DoT licence for the export of the aircraft on non-scheduled commercial or private flights abroad. The exception to this is when an aircraft is leaving only temporarily, on provision that the pilot produces a *Carnet de Passage en Douane*, or a Customs and Excise Form C42 in duplicate, in which a written undertaking is given that the aircraft will return within one month.

To facilitate the inwards clearance on reimportation of the aircraft, the pilot should retain the *Carnet de Passages en Douane* or the Customs and Excise Form C42 and present it to the Customs and Excise officer on return.

FLIGHTS BETWEEN THE CHANNEL ISLANDS AND GREAT BRITAIN AND NORTHERN IRELAND. For the purposes of arrival and departure procedures, the Channel Islands are regarded as being a place outside the UK (i.e. Great Britain and Northern Ireland).

FLIGHTS BETWEEN THE ISLE OF MAN AND GREAT BRITAIN AND NORTHERN IRELAND. All flights to and from the Isle of Man must land at Isle of Man/Ronaldsway Airport. Aircraft carrying goods from the Isle of Man into the UK or European Community must land at one of the Customs and Excise airports if duty has not been paid on these goods.

FORCED LANDINGS When an aircraft has been cleared by Customs and is outbound, or, while returning inbound from abroad it is forced to land at a place other than a Customs and Excise airport, the following procedures must be carried out:

1 The aircraft commander must report the landing immediately to a Customs and Excise officer or to a police constable.
2 He must produce the flight documents upon request by that officer or police constable.
3 He must ensure that no goods are unloaded and that the passengers and crew remain in the immediate vicinity, except when necessary for reasons of health, safety or the preservation of life or property.
4 He must comply with any directions given by that officer or police constable in reference to any goods carried.
5 If the aircraft is on an inward flight, the commander must make a report with regard to documentary requirements. This action may be deferred by the officer until the aircraft arrives at a Customs and Excise airport.

Aerodrome operating minima

Articles 32 and 32A of the ANO declare that aircraft registered outside the UK, both public and non-public transport, shall not fly over the UK unless the operator has supplied the crew with the aerodrome operating minima of all airfields of intended use.

A non-UK-registered aircraft, both public and non-public transport, shall not commence or continue an approach to landing at an aerodrome in the UK if the runway for the relevant runway and approach aid at the aerodrome is less than the operator's specified minimum, unless:

a The aircraft is below decision height; and
b The specified visual reference has been established at decision height and is maintained.

Apart from the crew being supplied and informed of this criteria, these instructions and prohibitions must also be submitted with the aerodrome operating minima to the CAA by the operator.

Flights to and from the UK

DOCUMENTARY REQUIREMENTS FOR SCHEDULED FLIGHTS

1 Arrivals
On a landing being made at a Customs and Excise airport, the aircraft commander must deliver the aircraft to the Examination Station and must provide the following documentation:

a A General Declaration (UK Form C155) if required by the Customs and Excise officer
b Two copies of the Cargo Manifest (UK Form C257) when cargo or unaccompanied baggage is carried
c Two copies of the stores list (on UK Form C209) if stores are on board
d At certain airports, the aircrew written personal declaration on Form C909.

2 Departures

a Notice of intended departure must be given to the Customs and Excise officer when the aircraft is brought to the Examination Station, and *before* any passengers, goods or stores are loaded aboard an aircraft departing for a place or area outside the UK
b The Customs and Excise officer requires two copies of the Cargo Manifest (UK Form C257) when cargo or unaccompanied baggage is to be loaded
c The Customs and Excise officer also requires two copies of the stores list on Form C208 for stores either remaining on board, or to be loaded on to the aircraft
d If, once Customs have been cleared on an outward journey, the aircraft lands at another specified Customs and Excise airport named in the outward clearance form and further stores and goods are loaded on board, Manifests and Stores Lists are required by paragraphs (**b**) and (**c**) above.

3 Transit

When a landing and departure at a UK Customs and Excise airport is made by an aircraft en route from one foreign country to another, without loading or unloading any cargo, mail, unaccompanied baggage or stores, a General Declaration is all that is required by the Customs and Excise officer on arrival, and two copies of the Cargo Manifest on departure. If any goods are loaded or unloaded while in transit, the full arrival and departure documentation must be produced to the Customs and Excise officer.

DOCUMENTATION FOR NON-SCHEDULED COMMERCIAL FLIGHTS All aircraft must comply with the regulations and restrictions governing all incoming flights from areas and places outside the UK, that the first landing be made at a Customs and Excise airport, and that the last take-off for all outgoing flights to areas and places outside the UK be made from a Customs and Excise airport.

Every operator is reminded that compliance with local restrictions on flying, noise abatement procedures, and noise characteristics as prescribed in the relevant pages of the UK AIP, AGA section, must be submitted and approved by the DoT.

Arrival and departure documentation is the same as for scheduled flights, added to export licensing requirement documentation as mentioned earlier in this chapter.

Advance arrangements must be made for the ground handling of aircraft, and to ensure that arrivals are scheduled during the airport's normal hours of watch, unless special arrangements have been made.

Cabotage

Aircraft registered in foreign countries which include the Commonwealth (but not Colonial registered aircraft), that are a party to the Chicago Convention, will not be granted permission to embark or disembark passengers and/or goods on a flight between any two places within the UK or in a UK Overseas territory or Associated State.

Arrival, departure and transit of passengers and crew

IMMIGRATION REQUIREMENTS

Passengers

For immigration purposes, the UK, the Channel Islands, the Isle of Man and the Republic of Ireland collectively form a common travel area. Passports are required by persons departing the UK to a place outside this common travel area, and persons arriving in the UK from outside the common travel area. The exceptions to this are nationalised persons of the European Community, who may enter the UK for visiting or employment provided they hold a national identity card. For employment purposes they also require a permit from the Department of Employment. Transitting passengers are not required to produce documents.

Powers and obligations of Captains and Owners or Agents under the Immigration Act 1971

- **a** Except with the prior approval of the Secretary of State, to embark and disembark passengers only at an airport designated as a port of entry, unless there is reasonable cause to believe them all to be citizens of the UK
- **b** To ensure that disembarkation does not take place by passengers until they have been examined by an Immigration Officer
- **c** To ensure that passengers are presented for examination in an orderly manner
- **d** To provide a list of crew and passengers if required
- **e** To ensure that all passengers who are non-UK citizens or European Community nationals are provided with a landing card
- **f** To remove or make arrangements for the removal from the UK, of an alien passenger who is refused entry to the UK
- **g** To detain in custody persons placed on board an aircraft under the authority of an Immigration Officer
- **h** To pay on demand by the Secretary of State any expenses incurred by any detention of persons, if required to do so
- **j** To remove a person against whom a deportation order is in force, if so directed by the Secretary of State.

Landing cards

Landing cards must be completed and produced to the Immigration Officer by all passengers over the age of 16 who are non-UK citizens or European Community nationals.

Aircrew

When persons arrive as members of the crew, and they are subject to the immigration control, they may enter the UK without leave if their engagement requires them to depart within seven days on the departure of their aircraft, unless:

a There is in force a deportation order against them; or
b They have at some time been refused leave to enter the UK and have not since been given leave to enter or remain in the UK; or
c An Immigration Officer requires them to submit to an examination.

An aircrew member who lawfully enters the UK and wishes to stay longer than the assigned seven days must seek permission from the Immigration Officer.

A valid crew licence or crew member's certificate which certifies the holder to re-enter the state of issuance is an acceptable document of identity.

Customs and Excise requirements

a Aircrew members are generally required to make a written declaration on Form C909
b Passengers may give an oral declaration which is accepted in respect of accompanied baggage presented for clearance by the owner.

Public health requirements

To prevent danger to public health, an examination and the necessary measures may be carried out by the Airport Medical Officer on the following persons:

a A person entering the UK who is suspected of suffering from, or to have been exposed to infection from, an infectious disease or is suspected of being verminous
b A person proposing to depart from the UK if there are reasonable grounds for believing that person to be suffering from a quarantinable disease subject to the International Health Regulations.

A valid international certificate for smallpox is required by persons entering the UK from certain countries, and if no certificate is presented, then a smallpox vaccination will be offered, before permission to enter, with a short term of surveillance or isolation of that person.

Cargo

IMPORTATION – Authority from a Customs and Excise officer is required before any goods can be unloaded from an aircraft, and they will be required to be unloaded only at the airport Examination Station. If an aircraft is unable to be taken to the Examination Station, the goods must be unloaded under the supervision of a Customs and Excise officer and taken to the Examination Station. Goods intended for onward transit may remain on board the aircraft by permission of the Customs and Excise officer. All imported goods unloaded must be deposited in the transit shed unless permission is given by the Customs and Excise officer.

EXPORTATION – Goods may not be loaded on an aircraft about to depart from a Customs and Excise airport on a flight for which outward clearance is required:

- **a** Until application for such clearance has been made or notice of intended departure has been given; or
- **b** At any place other than the Examination Station; or
- **c** Without the authority of the Customs and Excise officer.

The authority of the Customs and Excise officer is required if goods which have been loaded, or retained on board, for exportation or use as stores, are required to be unloaded.

TRANSIT – All cargo remaining on board for onward transit to a destination outside the UK should have the general statement – 'Part cargo remaining on board for exportation', written on the cargo manifest, and the cargo must not be interfered with while the aircraft is at the Customs and Excise airport.

CHAPTER 13
THE AIR NAVIGATION ORDER

The following are extracts of the main Articles and requirements of knowledge of the Air Navigation Order (CAP 393) which cover the pilot's examination for aviation law for the PPL and Professional Licences.

Aircraft to be registered (ARTICLE 3)

An aircraft flying in or over the UK must be registered in:

a Some part of the Commonwealth
b A contracting State
c Any other country which has in force an agreement between themselves and the Government of the UK.

The exceptions to this are aircraft that fall in to the following categories:

1 Gliders on flights which begin and end in the UK without passing over any other country, and also gliders that are not used for public transport or aerial work;

2 Any aircraft whose flight path begins and ends in the UK without passing over any other country, and which is flying in accordance with the 'B' Conditions set forth in Schedule 2 of the ANO, which briefly states that the aircraft shall fly only for the purpose of:

a Experimenting with or testing the aircraft and its equipment
b Allowing the qualification of a Certificate of Airworthiness to that aircraft
c Proceeding to or from a place whereby items (**a**) or (**b**) above can be carried out
d Demonstration of the aircraft for purposes of a sale of that or similar aircraft.

The qualification of registration as outlined above is not applicable to any kites or captive balloons.

Registration of aircraft in the UK (ARTICLE 4)

Article 4 of the ANO, Section 1, states many requirements for the registration of aircraft within the UK. The relevant paragraphs from Article 4 are set out in the following order against their appropriate paragraph reference number:

6 Application for the registration of an aircraft in the UK shall be made in writing to the Authority (CAA), and shall include or be accompanied by such particulars and evidence relating to the aircraft and the ownership and chartering therefore as it may require to enable the CAA to determine whether the aircraft may be properly registered in the UK and granted a certificate of registration. The application shall include a description of the aircraft as set out according to column 4 of the 'General Classification of Aircraft' laid down in Part A of Schedule 1 of the ANO, which briefly is as follows:

- **a** Free balloon, non-power-driven, lighter-than-air aircraft
- **b** Captive balloon, non-power-driven, lighter-than-air aircraft
- **c** Airship, power-driven, lighter-than-air aircraft
- **d** Glider, non-power-driven, heavier-than-air aircraft
- **e** Kite, non-power-driven, heavier-than-air aircraft
- **f** Aeroplane (landplane), power-driven, heavier-than-air aircraft
- **g** Aeroplane (seaplane), power-driven, heavier-than-air aircraft
- **h** Aeroplane (amphibian), power-driven, heavier-than-air aircraft
- **j** Aeroplane (self-launching motor glider), power-driven, heavier-than-air aircraft
- **k** Powered lift (tilt-rotor), power-driven, heavier-than-air aircraft
- **l** Rotorcraft (helicopter or gyroplane), power-driven, heavier-than-air aircraft.

10 If, once the aircraft has been registered in the UK, the ownership or part ownership of that aircraft is acquired by an unqualified person, either by legal entitlement or beneficial interest, the registration will become void and the certificate of registration must be returned to the CAA by the registered owner.

11 The CAA must be informed in writing by the registered owner of a UK registered aircraft when any of the following events occur:

- **a** Any change in the required particulars that were given to the CAA when the aircraft registration application was originally made
- **b** The aircraft's destruction or the permanent withdrawal from use
- **c** The termination of the demise charter that is given to a qualified person able to register the aircraft.

12 Any person who becomes the owner of an aircraft registered in the UK shall inform the CAA in writing to that effect within 28 days.

13 The aircraft registration may be amended or cancelled by the CAA when it appears necessary, and the registration cancellation shall take place within 2 months of the CAA becoming satisfied that a change of the aircraft's ownership has taken place.

Certificate of Airworthiness to be in force (ARTICLE 7)

1 An aircraft may fly within the UK if it has a duly issued Certificate of Airworthiness (C of A) valid under the law of the country of its registration, and compliance is made of any conditions to which the certificate was issued or validated.

The exceptions to the above requirements are flights beginning and ending in the UK without passing over any other country, of the following:

a A glider, not being used for aerial work nor the public transport of passengers
b A balloon, not being used for the public transport of passengers
c A kite
d An aircraft flying in accordance with the 'A' or 'B' Conditions set forth in Schedule 2 of the ANO
e An aircraft flying in accordance with the conditions of a Permit to Fly issued by the CAA in respect of that aircraft.

2 For UK registered aircraft, the C of A mentioned above must be issued or rendered valid in accordance with the requirements set out in Article 8 of the ANO.

Issue of Certificate of Airworthiness (ARTICLE 8)

2 An aircraft may only be flown for the purposes laid down in the categorisation requirements in Schedule 3 of the ANO as stated on the aircraft's C of A. The categories and purposes as specified in Schedule 3 are set out as follows, along with their period of validation:

a Transport category (Passenger)	Any purpose
b Transport category (Cargo)	Any purpose other than public transport of passengers
c Aerial work category	Any purpose other than public transport
d Private category	Any purpose other than public transport or aerial work

e Special category	Any purpose other than public transport, specified in the C of A but not including the carriage of passengers unless expressly permitted.

For categories (**a**) to (**d**) above, the validation periods are:
aircraft over 2,730 kg – 1 year; aircraft under 2,730 kg – 3 years
For category (**e**) above, the validation periods are:
aircraft over 2,730 kg – 1 year; aircraft under 2,730 kg – 1 year

Certificate of Maintenance Review (ARTICLE 9)

1 Section 1 of Article 9 of the ANO states that any aircraft registered in the United Kingdom which holds a C of A shall not fly unless:

a The aircraft is maintained to a maintenance schedule approved by the CAA. This maintenance schedule includes all the aircraft's equipment, which covers also the radios and particularly the engines
b There is in force a certificate, and that certificate of maintenance review shall certify the date when that maintenance review was carried out and also when the next review is due.

2 The purpose of issuing a certificate of maintenance review and the occasions when that review must be carried out shall be specified on the approved maintenance schedule as stated above.

3 A certificate of maintenance review can only be issued by:

a A licensed engineer of the UK, or under the law of any country which entitles that engineer to issue that certificate
b In a particular case when the certificate of maintenance is issued by a CAA authorised person
c When the CAA approves a person as being competent to issue such certificates.

4 Before the certificate of maintenance is issued, the authorised person/licensed engineer must verify that:

a The appropriate maintenance (as per the schedule),
b The inspections and modifications (as required),
c Rectifications of all the defects listed in the technical log, and
d The issuance of the certificate of release to service in accordance with the ANO,

have all been carried out.

5 If the aircraft is for public transport or aerial work, one copy of the certificate of maintenance shall be carried on the aircraft and the other copy is to be retained by the operator; a certificate of maintenance having to be issued in duplicate.

6 For a period of 2 years after its issue, a certificate of maintenance must be kept by the operator.

Technical log (ARTICLE 10)

1 An aircraft that is registered in the UK and holds a valid certificate of airworthiness in either the transport or aerial work category must have a technical log kept for it.

2 The commander of an aircraft that requires a technical log shall ensure after every flight that the following details be entered therein:

a The aircraft take-off and landing times
b Any defect, affecting the airworthiness or safe operation of the aircraft. In the event of no defects, an entry to that effect should be entered
c Any particulars relating to the airworthiness or operation of the aircraft, as required by the CAA.

The above details are to be kept in a CAA approved record for aircraft of which the maximum total weight does not exceed 2,730 kg and which is operated by a person who does not hold or require to hold an air operator's certificate. That person will be required to sign and date such entries.

Under certain conditions these entries may be made at the end of the flying day, unless a defect is detected during an earlier flight. These conditions covers consecutive flights which begin and end as follows:

a Within the same period of 24 hours
b At the same aerodrome
c The commander of the aircraft remains the same.

NOTE 1: With reference to item (**b**) above, the proviso excludes the dropping or projecting of any material for agricultural, public health or similar purposes.

NOTE 2: Technical logs or their CAA approved counterparts are required to be kept by the operator and preserved until a date 2 years after the aircraft has been permanently withdrawn from service, or any shorter period approved by the CAA.

Inspection, overhaul, repair, replacement and modification (ARTICLE 11)

1 A UK registered aircraft must not fly until it has been issued with a certificate of release to service after the aircraft, or a part of the aircraft, or any of the aircraft's equipment affecting the airworthiness of that aircraft, has been overhauled, repaired, replaced, modified, maintained or has been inspected according to Article 8(7) (b) of the ANO requirements.

The exception to this is when the repair or replacement to the aircraft or its equipment is actioned at a place that makes it impracticable to carry out the repair or replacement in a manner that a certificate of release to service can be issued, or it is not reasonably practicable for such a certificate to be issued while the aircraft is at that place.

In this instant the aircraft may fly to the nearest place where a certificate of release to service can be issued, if it is in the reasonable opinion of the commander that the aircraft can safely fly by a route for which it is properly equipped; and it is reasonable to fly to that destination while having due regard to any hazards that may affect the liberty or health of any person on board.

In the event of this flight having to take place, the commander is required to submit the written particulars of the flight, and the reasons for making it, to the CAA within 10 days of the flight having taken place.

2 An aircraft of which the maximum total weight authorised does not exceed 2,730 kg, and which holds a C of A in the special category, does not require a certificate of release to service unless the CAA requires one in a particular case.

5 A certificate of release to service shall be a certificate that qualifies that the aircraft, or any part of its equipment, has been overhauled, repaired, replaced, modified or maintained in complete compliance with the CAA directives as to the manner and with material approved by the CAA. With regards to inspections, it qualifies that the inspection of the aircraft or its equipment has been carried out in an approved manner and to a standard approved by the CAA.

6 (e) A certificate of release to service may be issued by the holder of an Airline Transport Pilot's Licence (Aeroplanes), a Senior Commercial Pilot's Licence (Aeroplanes) or a Flight Navigator's Licence granted or rendered under the ANO for any adjustments and compensation of direct reading magnetic compasses.

Equipment of aircraft (ARTICLE 13)

1 For an aircraft to fly legally it must be equipped in accordance with the law of the country in which it is registered. This requirement must enable lights and markings to be displayed, and signals to be made in accordance with this ANO regulation.

2 The equipment requirement is laid out in Schedule 4 of the ANO, for UK registered aircraft, and it must be of the type approved by the CAA and it must be installed in a manner approved by the CAA. Paragraph 3 of Schedule 4 lists equipment that does not require approval by the CAA. The contents of Schedule 4 are required to be known by both private and professional pilots. An example of this knowledge is given in the following extract from Schedule 4, Scales Y1 and Y2:

Scale Y1: If an aircraft has a total seating capacity of 60 but not more than 149 passengers, one portable battery-powered megaphone is required. However, if the total seating capacity is for 150 passengers or more, two megaphones are required. Please note that the inference is not on the number of passengers on board, but on the total seating capacity of that aircraft.

Scale Y2: With Scale Y2, the emphasis is different in that with the aircraft's C of A, if it carries more than 19 (20 plus) passengers and fewer than 100 passengers (99 or less), one portable battery-powered megaphone is required. The rest of Scale Y2 needs close study. However, the difference between Scale Y1 and Scale Y2 requirements is that Scale Y1 refers to total seating capacity, not to passengers. Scale Y2 refers to number of passengers on board and not to total seating capacity.

Radio equipment of aircraft (ARTICLE 14)

The schedule of radio equipment required is laid down in Schedule 5 of the ANO. This covers any aircraft flying within the UK and therefore aircraft of other countries must comply with the law of the country of registration, to enable communications to be made and the aircraft to be navigated, in accordance with the provisions of the ANO, Article 14.

Aircraft, engine and propeller log books (ARTICLE 16)

1 Apart from any other log book required to be kept by order of the ANO, UK registered aircraft must keep:
a An aircraft log book
b A separate log book for each engine fitted, and
c A separate log book for each variable-pitch propeller fitted to the aircraft.

The required particulars are set out in Schedule 6 of the ANO, and where the maximum total weight authorised is 2,730 kg or less, a CAA approved type of log book must be used.

2 (**a**) All of the log book entries should be made as soon as possible after an event has taken place, but it should not be more than 7 days after the expiration of the certificate of maintenance review. The exceptions to this are the entries referred to in Schedule 6, sub-paragraphs 2(d)(ii) and 3(d)(ii).

(**b**) The entries referred to in sub-paragraphs 2(d)(ii) and 3(d)(ii) state that the sum total of the times between take-off and landings for all flights made by that aircraft since the immediately preceding occasion that any maintenance, overhaul, repair, replacement, modification or inspection was undertaken on the engines or propellers must be entered.

5 The operator of the aircraft is required to keep the log book until a date 2 years after the engine or propeller has been destroyed or permanently withdrawn from use.

Aircraft weight schedule (ARTICLE 17)

1 Where a C of A has been issued as valid under the ANO, for any flying machine or glider, it must be weighed, and its centre of gravity determined in accordance with a manner approved by the CAA.

2 When the aircraft has been weighed, the aircraft operator must prepare a weight schedule indicating:

a Either the basic weight of the aircraft, or a weight approved by the CAA. (BASIC WEIGHT = aircraft weight empty + weight of unusable fuel and oil + equipment weight as indicated in schedule.)

b either the centre of gravity of the aircraft under basic weight conditions or a position of centre of gravity approved by the CAA.

3 The aircraft operator is required to preserve the weight schedule for a period of six months after the issue of a new weight schedule.

Aircrew Licensing

Composition of crew of aircraft (ARTICLE 19)

1 For any aircraft to fly it must carry a flight crew of the number and description as laid down by the law of the country in which the aircraft is registered.

For aircraft registered in the UK the following rules under ARTICLE 19 are applicable:

2 Adequate flight crew must be carried by all aircraft, both in amount of crew and crew description to ensure safety of that aircraft, and must be of at least the number required as specified on the C of A.

3 a A flying machine having a maximum total weight authorised in excess of 5,700 kg and flying for the purpose of public transport must carry at least two pilots as members of the flight crew.

b On or after 1 January 1990, where an aircraft commander is flying an aircraft of MTWA of 5,700 kg or less for public transport purposes under IFR, no fewer than two pilots must be carried if the aircraft is powered by:

- **i** One or more turbine jets
- **ii** One or more turbine propeller engines and has a pressurised personnel compartment
- **iii** Two or more turbine propeller engines and is allowed by its C of A to carry ten or more passengers
- **iv** Two or more turbine propeller engines and carries nine or fewer passengers in an unpressurised cabin, unless it is equipped with a serviceable CAA approved autopilot on take-off
- **v** Two or more piston engines, unless it is equipped with a serviceable CAA approved autopilot.

NOTE 1: Items (**iv**) and (**v**) above therefore allow only one pilot if a serviceable CAA approved autopilot is carried. If the autopilot is found to be unserviceable before take-off, it may still fly with one pilot if it is flown in accordance with a CAA approved arrangement.

NOTE 2: If the aircraft is flown in accordance with the terms of a police air operator's certificate, aircraft described in items (**iii**), (**iv**) or (**v**) above may fly with only one pilot.

4 If a public transport aircraft is intended to fly on a route or has a diversion, that is planned before take-off, for a distance of more than 500 nm from the take-off point, then a flight navigator or navigational equipment approved by the CAA must be carried when passing over or through any part of the specified areas in Schedule 7 of the ANO. Although the Schedule provides latitudes and longitudes (which do not have to be remembered), the basic areas should be known. These cover the following areas:

- **a** Arctic
- **b** Antarctic
- **c** Sahara
- **g** Indian Ocean
- **h** North Atlantic Ocean
- **j** South Atlantic Ocean

d	South America	**k**	Northern Canada
e	Pacific Ocean	**l**	Northern Asia
f	Australia	**m**	Southern Asia

NOTE: If a flight navigator is required to be carried, he/she must be in addition to any other crew member required to carry out other duties.

5 If an aircraft has radio communication apparatus which includes radiotelegraph (W/T) equipment, a flight radio operator is required as a member of that aircrew in addition to any other person required to carry out other duties, as required under Article 14 of the ANO.

6 In the interests of safety, the CAA may require an aircraft operator to carry additional persons as flight crew.

7 In the interests of safety on public transport flights, cabin attendants extra to the flight crew complement must be carried when 20 or more passengers are carried or when the C of A clears that aircraft to carry more than 35 passengers and on which at least one passenger is carried. On flights covered by these requirements at least one cabin attendant must be carried for every 50 passenger seats, or fraction of 50 passenger seats. Written permission may be granted by the CAA to carry fewer cabin attendants, provided that the operator complies with all the terms and conditions that the permission grants.

8 Additional cabin attendants may be required as laid down by the CAA directive to the operator in the interests of safety.

Members of flight crew – requirement of licences (ARTICLE 20)

1 A person must not act as a flight crew member on a UK registered aircraft unless that person holds an appropriate valid licence as laid down in the ANO. The exceptions to this allow a person who is a non-licence holder within the UK, the Channel Islands, and the Isle of Man to:

a Act as a flight radio-telephony operator while operating as a glider pilot (not flying for the purpose of public transport or aerial work); or as a person undergoing training in a UK registered aircraft as a flight crew member; or they are authorised by a licence holder of a radio-telephony station; or messages are transmitted only for instructional purposes, or for the safety of the aircraft or navigational purposes

b Act as pilot in command of an aircraft while being trained or tested for the grant or renewal of a licence by a qualified flying instructor or assistant flying instructor; and the person must be at least 17 years old and carry a valid medical certificate; no other person must be carried in these circumstances, and the aircraft must not be used for public transport or aerial work other than flying instruction

c Act as pilot of an aircraft for the purpose of qualifying as flight crew for the grant or renewal of a pilot's licence or rating; the aircraft must have dual controls and must not be used for public transport or aerial work other than for flying instruction or testing of the said pilot; the instructor must be in a position to use the controls

d Act as pilot-in-command of an aircraft at night if he/she holds an appropriate valid licence which does not include an instrument rating and he/she has not carried out in the previous 13 months as pilot in command at least 5 take-offs and landings at a time when the depression of the centre of the sun was 12 degrees or more below the horizon; he/she obeys the instructor who is qualified to give night instruction; the aircraft is not flying for public transport or aerial work other than that for instructional flying purposes

e Act as pilot-in-command of a balloon if he/she holds an appropriate valid licence but has not carried out 5 flights each of 5 minutes or more duration, as pilot-in-command, within the previous 13 months; he/she is under the instruction of a pilot authorised by the CAA to supervise flying in the type of balloon; the balloon is not flying for the purposes of public transport or aerial work other than that required for the flying instruction being given.

6 A non-licence holder may act as a member of the flight crew of a UK registered aircraft if acting in the course of their duty as a member of any of Her Majesty's naval, military or air forces.

Grant, renewal and effect of flight crew licences (ARTICLE 21)

1 The CAA will grant and issue licences for flight crew members to operate on UK registered aircraft in the classes laid down in Part A of Schedule 8 of the ANO once it is satisfied that each applicant is fit to hold that licence by virtue of being physically and mentally fit, after taking the appropriate medical examinations, and is qualified by reason of knowledge, experience, competence and skill, having taken an approved course of instruction and testing in the appropriate category; the applicants must also be of the minimum age or more for their crew categories; the licence will only be valid after it has been signed by the applicant (in ink) and shall last only for a period indicated on the said licence.

3 Part B of Schedule 8 lays out the different ratings that may be included on a licence by the CAA, when the applicant has satisfied the CAA that he/she has qualified in that capacity.

4 When a person has obtained a licence under the provisions of the ANO and in one of the classes laid down in Part A of Schedule 8, the holder of that licence will be entitled to perform the functions as specified in that Schedule. The specified functions as laid down in Part A are covered under the heading of 'Privileges'. Similarly, any rating obtained as part of that licence will entitle the holder to carry out the functions of that rating, as specified under Part B of Schedule 8.

5 Professional pilots' and flight engineers' licences must show that they hold a valid certificate of test or valid certificate of experience with reference to the functions they are allowed to perform, and also the type of aircraft rating to which it applies. Holders of private pilots' licences, however, who wish to exercise their privilege of an aircraft rating, must have the certificate of test or certificate of experience included in their personal flying log books.

6 For people to perform the functions of an Instrument Rating for aeroplanes or helicopters, IMC rating or flying instructor's or assistant flying instructor's rating, an appropriate valid certificate of test with reference to that rating must be included on their licences.

7 On the failure to pass the renewal test on the last occasion for which that test was taken, the licence then becomes invalid from that date of test, (i.e. if there is two months' validation still on the licence when the renewal test is taken and subsequently failed, that licence becomes immediately invalid, not at the end of the two months' validation).

8 Other than a flight radiotelephony operator's licence, the holder of any other flight licence must hold a valid medical certificate which is then deemed to be part of that licence. The CAA can require a licence applicant or licence holder to submit to a medical examination by a CAA approved person at any time it thinks fit.

9 When a person's physical or mental condition renders them unfit, either temporarily or permanently, their entitlement to act as a flight crew member of a UK registered aircraft becomes invalid.

This means that holders of medical certificates must inform the CAA in writing when they suffer any personal injury which prevents them carrying out their duties as flight crew members; or they suffer any illness lasting 20 days or more which prevents their functioning as flight crew members; or in the case of a woman, if she believes herself to be pregnant.

That person must inform the CAA in writing as soon as possible of injury or pregnancy, and in the case of illness, as soon as the 20 days have elapsed. His/her medical certificate will be suspended in all three cases. However, in the case of injury or illness, the suspension will cease after an approved medical examination has pronounced that person fit to resume their duties. In the case of pregnancy, the suspension may be lifted for a period of time and under conditions that the CAA think fit, and in any case after the pregnancy has ended and that person has been pronounced fit to resume duty as a flight crew member.

11 When the CAA thinks fit, it may grant an approval for a test to be carried out in a CAA approved flight simulator.

12 At the discretion of the CAA, it may approve any course of training or instruction, authorise a person to conduct such examinations or tests that it may specify, or approve a person to provide any course of training or instruction.

Validation of licences (ARTICLE 22)

The CAA may, at its own discretion, issue certificates of validation of any flight crew licences of aircraft which are granted under the law of any non-UK country. The periods and conditions of these validations are at the discretion of the CAA.

Personal flying log books (ARTICLE 23)

1 All flight crew members, and all those who fly for the purpose of qualifying for the grant or renewal of a licence, with reference to UK registered aircraft, must keep a personal flying log book, within which is recorded:

a The name and address of the holder of the log book
b The flight crew member's licence particulars
c The employers name and address (if any).

2 The member of an aircraft's flight crew, or those persons flying to qualify for a grant or renewal of a licence, must enter the particulars of each flight immediately or as soon as possible after the end of each flight. These particulars will include:

a The date and the places of embarkation and disembarkation of each flight, including the time spent acting in their capacity as a flight crew member or performing their qualifying duties
b The aircraft type and registration marks
c The capacity in which the holder acted in flight

d Particulars of any special conditions under which the flight was conducted, including night flying and instrument flying
e Particulars of any test or examination undertaken whilst in flight.

3 A helicopter shall be deemed to be in flight from the moment the helicopter first moves under its own power for the purpose of taking off until the rotors are next stopped.

4 Particulars of any test or examination whilst in a flight simulator shall be recorded in the log book, including:

a The date of the test or examination
b The type of simulator
c The capacity in which the holder acted
d The nature of the test or examination.

Instruction in flying (ARTICLE 24)

1 A person shall not give any flying instruction unless he/she holds a licence entitling the pilot to act as pilot-in-command, and which includes a flying instructor's rating or assistant flying instructor's rating.

2 This article applies to flying instruction on a flying machine or glider for the granting of a pilot's licence or the inclusion or variation of any rating.

This Article does not apply to previous title holders or those qualified in Her Majesty's naval, military or air forces to act as pilots of multi-engined aircraft.

Operations Manual (ARTICLE 26)

1 An Operations Manual must be made available for all UK registered aircraft that are used for the purpose of public transport, except where the flights are not intended to exceed 60 min and are either flights for training or are intended to begin and end at the same aerodrome. Operators of aircraft involved solely for use under the terms of a police air operator's certificate are exempt from this article.

2 An Operations Manual must be kept up to date by the operator and be made available to each member of the operating staff, and the relevant parts affecting crew duty responsibilities must be accessible to every flight crew member during each flight. All information and instructions in the Operations Manual must contain all specified matters in Part A of Schedule 10 of the ANO, and must be sufficient for all operating staff to perform their duties correctly.

NOTE: Information and instructions available in the Flight Manual need not be repeated in the Operations Manual.

Training Manual (ARTICLE 27)

1 Except for an aircraft flying, or intended to fly, under the terms of a police air operator's certificate, every operator of a UK registered aircraft used for the purpose of public transport must ensure that an up-to-date training manual is kept and made available to every person appointed by the operator to give or to supervise the training, experience, practice or periodical tests required by the ANO.

Public transport – operator's responsibilities (ARTICLE 28)

1 Before the operator of a UK registered aircraft allows it to fly for the purpose of public transport, he/she must ensure for that flight that a pilot from among the flight crew is designated as aircraft commander; the safe navigation of the aircraft is ensured by satisfying themselves that there are adequate aeronautical radio stations and navigational aids serving the intended route; the safety of the aircraft and passengers are ensured by satisfying themselves that the intended aerodromes and alternates for landing are adequately manned and equipped, making their purpose suitable for the intended landing.

NOTE: The aircraft operator will not be required to be satisfied to the adequacy of fire-fighting, search, rescue or other services in the event of an accident.

2 The operator of a UK registered aircraft must maintain, preserve, and when necessary produce and furnish records as to the training, experience, practice and periodical tests of any person allowed to fly as a flight crew member in his/her aircraft for the purpose of public transport, and to ensure that that person is qualified to do so, to the standard specified in Part B of Schedule 10 of the ANO. The exceptions to this are those persons who are undergoing training for the performance of these duties.

3 In a UK registered aircraft used for the purpose of the public transport of passengers, the operator must not permit any member of the flight crew to simulate emergency manoeuvres or procedures which will adversely affect the flight characteristics of the aircraft.

Public transport – operating conditions (ARTICLE 30)

1 Except for the purpose of training, a UK registered aircraft shall not fly for public transport purposes unless compliance is made with the requirements with reference to its weight and related performance, and flight in specified meteorological conditions or at night.

2 The information as to its performance (contained in the C of A) relating to that aircraft, or the best available information that the aircraft commander has, will form the basis of the assessment of the aircraft's ability to comply with paragraph **(1)** above.

3 When flying over water, a UK registered flying machine used for public transport purposes must fly at such an altitude that would enable it to reach a place and land safely in the event of failure of:

a Its only engine (one-engined aircraft)
b One engine with the remaining engines operating at maximum continuous power (multi-engined aircraft).

The exceptions to this are aircraft taking off and landing.

United Kingdom registered aircraft – aircraft operating minima (ARTICLE 31)

1 Article 31 applies to all UK registered aircraft used for public transport purposes.

1A **a** Particulars of the aerodrome operating minima for every aerodrome of intended departure or landing and every alternate aerodrome must be established and included in the operations manual by the aircraft operator. In the circumstances where an operations manual of any type is not required, the operator must supply in writing (before the commencement of the flight) to the aircraft commander the particulars of the required data and instructions enabling the aerodrome operating minima to be calculated. A copy of this data and instruction must be kept by the operator, outside of the aircraft, for a minimum period of three months.

b Every aircraft operator who is required to have an operations manual must include in that operations manual, data and instructions to enable the aircraft commander to calculate the aerodrome operating minima for aerodromes where, before the beginning of each flight, the aircraft operator could not have reasonably foreseen this information.

2 Unless the CAA otherwise permits it in writing, the conditions at any aerodrome must not be less than the aerodrome operating minima specified to allow a safe landing or take-off.

3 An aircraft operator must take into account, when establishing aerodrome operating minima, the following:

a Any conditions in an aircraft's C of A, and also the type, performance and handling characteristics of that aircraft
b The crew composition
c The aerodrome and its surrounding physical characteristics
d The selected runway dimensions
e The relevant aerodrome aids in use (visual and otherwise) for approach, landing and take-off, of which the crew are trained in using; the procedures to be taken when using such aids or in the absence of them.

4 When Article 31 applies to an aircraft, it must not commence a flight at a time when:

a The departure aerodrome's cloud ceiling or runway visual range is less than the relevant minimum specified for take-off; or
b the information available to the aircraft commander indicates that a landing at the destination and alternate aerodromes would be impossible under the current prevailing conditions at the ETA.

5 An aircraft making a descent to an aerodrome must not descend from a height of 1,000 ft or more above the aerodrome to a height less than 1,000 ft above the aerodrome if the RVR is less than the specified minimum for landing at that aerodrome.

6 When making a descent to an aerodrome, an aircraft must not continue an approach to landing at any aerodrome by flying below the relevant specified decision height or below the relevant specified minimum descent height unless the specified visual reference for landing can be established and maintained from that height.

Public transport aircraft not registered in the UK – aerodrome operating minima (ARTICLE 32)

1 Public transport aircraft registered outside the UK must not fly in or over the UK unless the operator has supplied the CAA with particulars concerning the aerodrome operating minima at aerodromes within the UK. Any amendments and additions that the CAA requires to be made must be included in the aerodrome operating minima submitted by the operator before the operator's aircraft can fly in or over the UK.

3 A public transport aircraft not registered in the UK, and making a descent to an aerodrome in the UK, must not descend from a height of 1,000 ft or more above the aerodrome to a height less than 1,000 ft above the aerodrome if the RVR is less than the specified minimum for landing at that aerodrome.

4 When making a descent to an aerodrome, a public transport aircraft not registered in the UK must not continue an approach to landing at any aerodrome in the UK by flying below the relevant specified decision height or below the relevant specified minimum descent height unless the specified visual reference for landing can be established and maintained from that height.

Non-public transport aircraft – aerodrome operating minima (ARTICLE 32A)

1 Article 32A shall apply to any non-public transport aircraft.

2 An aircraft to which Article 32A applies, when making a descent at an aerodrome to a runway which has a notified instrument approach procedure, must not descend from a height of 1,000 ft or more above the aerodrome to a height less than 1,000 ft above the aerodrome if the RVR for that runway is less than the specified minimum for landing.

3 When making a descent to a runway which has a notified instrument approach procedure, an aircraft to which Article 32A applies must not continue an approach to landing on such a runway by flying below the relevant specified decision height or below the relevant specified minimum descent height unless the specified visual reference for landing can be established and maintained from that height.

Pilots to remain at controls (ARTICLE 33)

1 The commander of a UK registered flying machine or glider must ensure that one pilot remains at the controls during flight, and if the aircraft is required to carry two pilots, the commander must also ensure that they both remain at the controls during take-off and landing. If the flight is for public transport purposes and it carries two or more pilots, the commander himself must remain at the controls during take-off and landing.

2 While seated at the controls, all pilots must wear safety belts, with or without one diagonal shoulder strap, or a safety harness.

Pre-flight action by commander of an aircraft (ARTICLE 35)

The commander of a UK registered aircraft must satisfy himself before the aircraft takes off that:

- **a** The flight can safely be made, taking into consideration the route and aerodrome to be used, weather reports and forecasts, and any alternative plan of action that may be required owing to unforeseen circumstances
- **b** That all the equipment to be carried for the flight is in a fit condition; and that the conditions are being obeyed when an operator is granted permission under the minimum equipment requirements
- **c** The aircraft is fit for the intended flight and, where requirement of a certificate of maintenance review is needed, it will not expire during that flight
- **d** That the load can be safely carried for the intended flight with consideration to weight, distribution and security
- **e** That sufficient fuel, oil and coolant (where necessary) is carried, and an allowance made for contingencies. Compliance with instructions is made of the operations manual for a flight for the purpose of public transport
- **f** Sufficient ballast is carried for the flight in the case of airships and balloons
- **g** Having regard to the performance in the case of a flying machine, that it can safely take off, reach and maintain a safe height, and make a safe landing at the intended destination, having regard for any obstructions during these manoeuvres
- **h** Each member of the crew has complied with the pre-flight check system established by the operator and set down in the operations manual or elsewhere.

Passenger briefing by commander (ARTICLE 36)

Before a UK registered aircraft takes off on any flight, the commander must ensure that all the passengers are aware of the position and method of use of emergency exits, safety belts, safety harnesses, oxygen equipment (if required), lifejackets, floor path lighting systems and all other safety devices required to be used individually by passengers if an emergency occurs; all passengers must be instructed also in the action to take in an emergency.

Public transport of passengers – additional duties of commander (ARTICLE 37)

1 Article 37 applies to UK registered aircraft on flights for the public transport of passengers.

2 For every flight under which Article 37 is applicable, the aircraft commander must ensure that:

a If the aircraft is a landplane but during the course of its flight will be at a point more than 30 min from the nearest land, a demonstration in the use of lifejackets is given to all passengers; this demonstration must also be given if, when carrying cabin attendants, the aircraft is due to proceed beyond gliding distance of land, or, if an emergency occurs during take-off or landing, the aircraft is likely to land in water

b If the aircraft is a seaplane a demonstration of the use of lifejackets is given to passengers before take-off

c The crew are properly secured in their seats before the aircraft takes off; and cabin attendants (if carried) are secured in their seats in the passenger compartment, to assist passengers, before take-off

d Before the aircraft starts to move after embarkation of passengers and until after take-off, before landing until the aircraft has come to rest, and during turbulent air conditions or emergencies, the passengers of 2 years of age or more are secured in their seats by seat belts or harnesses, and those under 2 years of age by means of a child restraint device; that items of baggage in the passenger compartment, due to their size, weight or nature of the item, are properly secured, and in the case of an aircraft's seating capacity being more than 30 passengers, they must be stowed in approved stowage spaces in the passenger compartment

e In all aircraft with a C of A first issued (UK or abroad) on or after 1st January 1989, except where a pressure greater than 700 mb is maintained (i.e. pressurised aircraft) in all passenger and crew compartments throughout the flight, that before the aircraft reaches flight level 100 the method and use of oxygen is demonstrated to all passengers; that during any period the aircraft is flying above flight level 100 all the flight crew are on oxygen, and when above flight level 120 the passengers and cabin attendants are recommended to use oxygen

f In all aircraft with a C of A first issued (UK or abroad) before 1st January 1989, except where a pressure greater than 700 mb is maintained in all passenger and crew compartments throughout the flight, that before the aircraft reaches flight level 130 the method and use of oxygen is demonstrated to all passengers; that during any period the aircraft is flying above flight level 100 all the flight crew are on oxygen, and when above flight level 130 the passengers and cabin attendants are recommended to use oxygen.

Operation of radio in aircraft (ARTICLE 38)

1 The aircraft's radio station must only be operated under the conditions of a licence issued by the country of aircraft registration and by a duly licensed person or otherwise permitted under that law.

2 Whenever there is a requirement for an aircraft in flight to carry radio communications equipment, a continuous radio watch must be maintained by a flight crew member listening on a notified or designated frequency. A radio message may permit the radio watch to be discontinued or continued on another frequency. A radio watch may also be discontinued if a device (Selcal) is installed in the aircraft and the aeronautical radio station has been informed and granted permission, or, if the station is outside of the UK, they are able to transmit a suitable signal for this purpose.

3 Whenever the appropriate ATC has instructed the manner of use of the equipment required to be carried by an aircraft, a flight crew member shall operate the radio or radio navigation equipment accordingly.

4 During operation of an aircraft's radio station, it must not cause interference, and must only be operated for sending signals of such public correspondence permitted under the licence, and in accordance with general aeronautical international practice, only emissions of a class and frequency appropriate to the airspace in which the aircraft is flying; and distress, urgency and safety messages and signals; and messages and signals relating to the flight of the aircraft.

5 Every UK registered aircraft equipped with radio communications equipment must carry and keep a telecommunications log book containing information on:

a The aircraft radio station identification
b The date and time of the start and finish of every radio watch and frequencies maintained in the aircraft
c The date and time, with particulars, of all messages and signals received or despatched, especially of any distress signals or messages received or sent
d Action taken upon receipt of a distress signal or message
e Any communication failure or disruption and the cause if known.

7 The aircraft operator must preserve the telecommunication log book until a date six months after the date of the last entry.

8 When a UK registered flying machine is in controlled airspace below flight level 150, or is taking off or landing, and is being used for public transport purposes, the pilot and flight engineer must not use a hand-held microphone for radio communication nor for purposes of intercommunication within the aircraft.

Use of flight recording systems and preservation of records (ARTICLE 40)

1 A cockpit voice recorder or a flight data recorder or a combined cockpit voice recorder/flight data recorder must always be in use from the beginning of the take-off run to the end of the landing run for any flight that an aircraft is required to carry one under Schedule 4 of the ANO.

2 The aircraft operator is responsible for preserving the last 25 hours of recording made by any flight data recording system that is required to be carried and also of at least one representative flight made within the last 12 months which includes a take-off, climb, cruise, descent, approach to landing and a landing; this representative recorded flight must be able to be identified for the flight to which it relates. These records must be preserved for any period that is directed by the CAA.

3 On any helicopter flight that a cockpit voice recorder or a flight data recorder or a combined cockpit voice recorder/flight data recorder is to be carried, it must be in use from the time the first rotor turn for take-off purposes until the rotors are next stopped.

4 Helicopter operators must preserve the last 8 hours' recording of a flight data recording, but for a combined cockpit recorder/flight data recorder the last 8 hours of recording or the last 5 hours of recording of the last flight, whichever is the greater or for any period as the CAA may permit.

Towing of gliders (ARTICLE 41)

1 A glider may not be towed by an aircraft unless the C of A of an aircraft issued by the law of the country of registration permits it to do so.

2 A length of 150 m must not be exceeded by the total combination of aircraft, tow rope and glider in flight.

3 On an aircraft about to tow a glider, the commander of the aircraft must be satisfied before take-off that:

a The condition and strength of the tow rope is sufficient for the intended purpose; that the combination is capable of safely taking off, reaching, and maintaining a safe height for safe separation, having regards to performance, conditions and obstructions on the intended route, and that the towing aircraft can make a safe landing at the intended destination

b A signals system has been agreed with suitably stationed personnel for the glider to be able to take-off safely

c The towing aircraft commander and the glider commander have agreed emergency signals to be used by the towing aircraft commander to indicate immediate release of the tow by the glider, and by the glider commander to indicate that the tow cannot be released.

Towing, picking up and raising of persons and articles (ARTICLE 42)

1 Unless a provision in the C of A issued or rendered valid for an aircraft under the law of the country of registration permits it, an aircraft may not tow any article by external means, other than a glider, and it may not pick up or raise any person, animal or article.

2 The launching or picking up of tow ropes, banners or similar articles may be accomplished by an aircraft only on an aerodrome.

3 The towing of any article, other than a glider, may not be done at night or when flight visibility is less than one nautical mile.

4 A length of 150 m may not be exceeded by the combined length of the towing aircraft, tow rope and the article in tow.

5 When an article, person or animal is suspended from a helicopter, the helicopter may not fly at any height over any congested area of a city, town or settlement.

6 Unless a passenger is to be lowered or picked up by external means, or unless the passenger has duties in connection with an article, person or animal that is suspended from a helicopter, no passengers must be carried by that helicopter.

7 Article 42 does not prohibit the towing of any radio aerial, any instrument for experimental purposes, or any signal, apparatus or article by an aircraft in flight that has permission under the ANO; nor the picking up or raising of any person, animal or article in an emergency or for the purpose of saving life.

Dropping of articles and animals (ARTICLE 43)

1 The dropping of articles or animals (with or without a parachute) will not be permitted from an aircraft in flight which might endanger persons or property.

2 Aircraft flying over the UK shall not be allowed to drop any articles or animals (with or without a parachute) unless under terms of a police air operator's certificate or an aerial application certificate except that the commander of an aircraft may authorise the dropping of articles in the following circumstances:

a Articles for life-saving purposes
b The jettisoning, in case of emergency, of fuel or other articles in the aircraft
c Ballast in the form of fine sand or water
d Articles solely for the purpose of navigating the aircraft in accordance with ordinary practice or with the provision of the ANO
e The dropping at an aerodrome of tow ropes, banners, or similar articles towed by aircraft
f Articles with the permission of the CAA under set conditions for the purpose of public health or as a measure against weather conditions, surface icing or oil pollution, or for the training relating to these conditions
g By permission of the CAA under set conditions of the dropping of wind drift indicators for parachute descents.

3 For the purpose of Article 43, dropping includes projecting and lowering.

Carriage of weapons and of munitions of war (ARTICLE 46)

1 Munitions of war may be carried in an aircraft with the written permission of, and under conditions set by, the CAA. The aircraft operator has to inform, in writing, the commander of the aircraft (unless flown in accordance with a police air operator's certificate) before the flight commences of the weight, type, quantity and location of that munition of war.

2 Except for aircraft under terms of a police air operator's certificate, it is unlawful for an aircraft to carry any weapons or munitions of war in any compartment accessible to passengers.

3 For a person to carry, possess, take or cause to be taken a weapon or munition of war aboard an aircraft is against the law, unless it is either part of passenger baggage or consigned to be carried as cargo, and is carried in part of the aircraft that has no passenger access, and is unloaded in the case of a firearm. The consignor or the passenger concerned must furnish the operator with its particulars before the flight commences. The operator must also give consent for the items to be carried.

4 This article does not apply to non-UK registered aircraft whereby the law of that country permits legal carriage of that weapon or munitions of war to ensure the safety of the aircraft or its passengers.

5 For the purposes of Article 46, 'munitions of war' is defined as any weapon, ammunition or article containing an explosive or any noxious liquid, gas or other thing which is designed or made for use in warfare or against persons, including parts, whether components or accessories, for such weapon, ammunition or article.

Carriage of dangerous goods (ARTICLE 47)

1 The Secretary of State is responsible for making the regulations prescribing how certain articles and substances are classified as dangerous goods and which categories an aircraft may or may not carry. He is further responsible for prescribing the conditions applicable to the loading, suspension beneath and carriage by an aircraft, as well as the manner of packing, labelling and how they must be consigned beforehand. The documentation required, the persons who need to be informed and the provisions covering the safety of aircraft and its equipment, and the safety of persons and property over which the carriage of dangerous goods are carried, also come under his jurisdiction.

Method of carriage of persons (ARTICLE 48)

When an aircraft is in flight, a person may only be carried in suitable designed accommodation and shall not be carried on the wings or in the undercarriage. Temporary access to any part of the aircraft may be given to a person for taking necessary action to ensure the safety of the aircraft or of any person, animal or goods on board, or to check the cargo or stores carried in a suitably designed accessible compartment.

Exits and break-in markings (ARTICLE 49)

1 Article 49 applies to every UK registered public transport aircraft.

2 When an aircraft subject to Article 49 is carrying passengers, every exit and internal door of that aircraft must be in working order, and also must be kept free from obstruction during take-offs, landings and during emergencies and must not be locked or secured in any way that may prevent, hinder or delay passengers when exiting. If an exit or door is not required to be used by passengers for any reason, an exit may be obstructed by cargo if approved by the CAA. Internal doors may be allowed to be in non-working order provided they are placed in a manner that does not prevent, hinder or delay the exit of passengers in an emergency. A door between the flight crew compartment and any other compartment in the aircraft may be locked or bolted, at the aircraft commander's discretion, to prevent access by passengers to the flight crew compartment.

3 The words 'EXIT' or 'EMERGENCY EXIT', in capital letters, must be marked at every exit as appropriate to its use.

4 Instructions to indicate the correct method of opening an exit must be displayed in English, accompanied by a diagram. These instructions must be marked on or near the inside surface of the door or exit, and if it is openable from the outside they must also be marked on or near the exterior surface.

5 Every aircraft, except helicopters, having a MTWA in excess of 3,600 kg, must be marked externally on the fuselage indicating 'break-in areas' which allow easier external emergency access by persons outside the aircraft effecting a rescue by external sources.

Right-angled corner markings, with arms of 10 cm long and 2.5 cm wide, shall mark and indicate the rectangular break-in areas, along with the words 'CUT HERE IN EMERGENCY', in capital letters, across the centre of the break-in area.

6 All exits intended for use by passengers in an emergency on aircraft with a MTWA exceeding 5,700 kg must be marked on the aircraft exterior by a band not less than 5 cm wide outlining the exit.

7 The aircraft exit markings must be painted or fixed in a permanent manner, the colour to be in red, except where the background colour makes the red markings unclear, whereby the colour must be in white or some other contrasting colour. These markings must be kept clean, clear and unobscured at all times.

8 If the aircraft has one unserviceable exit only and is at a place where it cannot get the exit repaired or replaced it may fly to, and land at a place where it can have the exit repaired or replaced provided that the number of passengers carried and position of the seats occupied are approved by the CAA. Once approved the exit must be fastened by locking or otherwise, the words 'EXIT' or 'EMERGENCY EXIT' should be covered, and a red disc, of 23 cm or more diameter, with a horizontal white bar displaying the written words 'No Exit' in red letters, must mark that exit.

Drunkenness in aircraft (ARTICLE 52)

1 A person shall not enter any aircraft when drunk, or be drunk in any aircraft.

2 Members of the crew whilst acting in their aircrew capacity, or being carried for that purpose, must not be under the influence of drink or drugs whereby their ability to act as crew members is impaired.

Smoking in aircraft (ARTICLE 53)

Every UK registered aircraft must exhibit a notice, visible from every passenger seat, indicating when smoking is prohibited. Smoking is prohibited in any compartment that displays a 'No Smoking' sign when displayed by or on behalf of the aircraft commander.

Authority of aircraft commander (ARTICLE 54)

For the purpose of securing the aircraft's safety and that of persons or property carried therein, or the safety, efficiency or regularity of air navigation, all lawful commands given by the aircraft commander must be obeyed by all persons in a UK registered aircraft.

Stowaways (ARTICLE 55)

A person is not allowed to stow away on an aircraft unless he/she has the consent by a person entitled to give consent.

Fatigue of crew

Application and interpretation (ARTICLE 57)

1 The following two articles (58 & 59) of the ANO are to be applied to all UK registered aircraft engaged on flights for the purpose of public transport or operated by an air transport undertaking, but do not apply to flights made for the instruction of flying by or on behalf of a flying school or flying club.

2 The meanings of 'flight time' and 'day' given in this Part VI of the ANO shall have the following meanings:

a 'flight time' relates to the time spent in an aircraft, while in flight, by a crew member, regardless of the aircraft's country of registration (but does not apply to an aircraft with a MTWA of 1,600 kg or less and which is not flying for public transport or aerial work purposes)
b 'day' – beginning at midnight Co-ordinated Universal Time (UTC) for a period of 24 hours.

3 A helicopter is classified as being in flight from the moment it moves by its own power (for the purpose of taking off) until the rotors are next stopped.

Fatigue of crew – operator's responsibilities (ARTICLE 58)

1 An aircraft operator, as defined in Article 57, and to which Article 58 applies, must not allow a flight to be made by his/her aircraft unless:

a All persons flying in that aircraft as members of crew have their flight times regulated under an established scheme
b The scheme is approved and subject to conditions set out by the CAA
c The scheme is incorporated in the operations manual or in a document, where no operation manual is required, in which case all persons acting as members of the crew must have access to a copy
d All reasonably practicable steps are taken to allow all persons acting as crew members of that aircraft to comply with the scheme.

2 The aircraft operator must not allow a person to fly as a crew member if the operator knows or has reason to believe that the person could suffer fatigue to a degree that might endanger the safety of that aircraft or its occupants.

3 The aircraft operator must not allow a person to fly as a member of flight crew unless the operator possesses an accurate and up-to-date record of that person and of the 28 days preceding that flight, showing all his/her flight times and brief particulars of the duties performed by that person during the flight times recorded.

4 The aircraft operator must preserve the records referred to in paragraph 3 above for a period of 12 months after that flight.

Fatigue of crew – crews' responsibilities (ARTICLE 59)

1 Persons must not act as aircraft crew members if they know or suspect that they are suffering from or may suffer from such fatigue that may endanger the safety of the aircraft or its occupants.

2 All flight crew members must ensure that they do not fly in their flight crew roles unless they have ensured that the aircraft operator has or is aware of their flight times during the 28-day period preceding the flight.

Flight times – responsibilities of flight crew (ARTICLE 60)

A person shall not act as a flight crew member of a UK registered aircraft if at the beginning of the flight if the sum total of all that person's previous flight times:

a Exceed 100 hours during the preceding 28 consecutive days expiring at the end of the day that the flight begins; or

b Exceeds 900 hours during the period of 12 months expiring at the end of the previous month:

Except for flights made:

i In an aircraft of which the MTWA does not exceed 1,600 kg and is not flying for public transport or aerial work purposes; or

ii In an aircraft flying for non-public transport purposes nor operated by an air transport undertaking if at the beginning of that flight the sum total flying hours of that person does not exceed 25 hours since his/her last approved medical examination.

Documents to be carried (ARTICLE 61)

1 An aircraft may only fly if it carries the documents required under the law of the country of registration.

2 When in flight, a UK registered aircraft must carry the documents listed in Schedule 11 of the ANO. The exceptions to this are aircraft which begin and end flights at the same aerodrome and fly only in the UK airspace, in which cases the documents may be kept at the aerodrome.

Documents required in Schedule 11 are as follows:

i *Flight for public transport purposes*

a Aircraft radio licence and telecommunication log book
b C of A to include the Flight Manual
(**NOTE**: If the operations manual contains limitation and emergency procedures and also performance data, the flight manual need not be carried if written permission is given by the CAA)
c Flight crew licences
d One copy of the load sheet
e One copy of the certificate of maintenance review
f Technical log
g Operations Manual

and if the flight is international air navigation the following must be included:

h Certificate of registration
j Copy of procedures and signals between interception and intercepted aircraft.

ii *Flight for aerial work purposes*

a Aircraft radio licence and telecommunication log book
b C of A as stated in (**i**) above
c Flight crew licences
d One copy of the certificate of maintenance review
e Technical log

and if the flight is international air navigation, the following must be included:

f Certificate of registration
g Copy of procedures and signals between interception and intercepted aircraft.

iii *International air navigation private flight*
a Aircraft radio licence and telecommunication log book
b C of A as stated in (**i**) above
c Flight crew licences;
d Certificate of registration
e Copy of procedures and signals between interception and intercepted aircraft.

NOTE: The requirements regarding the carriage of the operations manual are set out in Article 26 of the ANO, as stated under that heading earlier in this chapter.

Production of documents and records (ARTICLE 63)

1 When an authorised person requests documentation to be produced, the aircraft commander must produce the following documents within a reasonable time:

a The certificates of registration and airworthiness
b The flight crew licences
c Any other document listed and required to be carried under Article 61 of the ANO.

3 A licence holder must produce his/her licence, which is to include any certificate of validation, and where required a medical certificate, to a person duly authorised and making a request for such documentation, within 5 days. The exception to this are licences required to be carried by an aircraft or kept at the aerodrome.

4 Every person required by the ANO to keep a personal flying log book must produce it to any authorised person upon their request for it within a reasonable time for up to 2 years after the date of the last entry made in it.

Offences in relation to documents and records (ARTICLE 68)

1 A person shall not with intent to deceive:

a Use any certificate, licence approval, permission, exemption or other document issued or required by the ANO which has been forged, altered, revoked or suspended, or to which they are not entitled

b Lend any certificate, licence approval, permission, exemption or other document issued or having effect or required by the ANO, or allow it to be used by any other person; or

c Falsely represent with intent to procure for theirself or any other person the grant, issue, renewal or variation of any such certificate, licence, approval, permission or exemption or other document.

Any reference to a certificate, licence, approval, permission or exemption or other document includes a copy or purported copy thereof.

2 During the required period of preservation, as stated in the ANO, of any log book or other record, a person must not intentionally damage, destroy, alter or render illegible any entry made therein, or knowingly make, or procure or assist in the making of any false entry in or material omission from any such log book or record.

3 The use of ink or indelible pencil is required in making all written entries in a log book or record.

4 A person shall not knowingly make any incorrect entry in any material particular, or any material omission in a load sheet.

Rules of the air (ARTICLE 69)

A separate section in the ANO gives all the statutory requirements under the heading of Rules of the Air Regulations, and these apply to:

a All aircraft within the UK, and with reference to low flying, also in the neighbourhood of an offshore installation; and

b All UK registered aircraft, wherever they may be.

2 It shall be an offence to contravene, to permit the contravention of, or to fail to comply with, the Rules of the Air. Lawful exceptions are laid out in the following paragraph (**3**).

3 Departure from the Rules of the Air will be deemed lawful to the extent necessary:

a For avoiding immediate danger

b For compliance with the law of any non-UK country in which the aircraft is flying; or

c For compliance with Military Flying Regulations or Flying Orders to Contractors issued by the Secretary of State in relation to an aircraft of which the commander is acting as such in the course of his duty as a member of any of HM Armed Forces.

4 If the Rules of the Air are departed from to avoid immediate danger, the aircraft commander shall supply written particulars of the incidence within 10 days, to the competent authority of the country in which the incident took place or to the CAA when the incident has taken place over the high seas.

5 The Rules of the Air do not allow provision for exoneration to any person for any consequence caused by their neglect in the use of signals or lights or of any neglect of precautions in any circumstances.

Power to prohibit or restrict flying (ARTICLE 74)

1 The Secretary of State may impose conditions or make regulations prohibiting or restricting flying by any UK or non-UK registered aircraft in any airspace over the UK or in the neighbourhood of an offshore installation, and by any UK registered aircraft in any other airspace where HM's Government provides navigational services. These regulations may apply generally or to particular types of aircraft. Furthermore, where it is deemed necessary the restriction or prohibition to flying may be made by reason of:

a The intended gathering or movement of a large number of persons
b The intended holding of an aircraft race or contest or an exhibition of flying; or
c National defence or any other reason affecting the public interest.

NOTE: A yellow Aeronautical Information Circular lays down the requirements for notification of unusual aerial activities, which includes the length of notice required for licensed and unlicensed aerodromes to hold various activities such as air shows, air races and other aerial competitions. This basically states the following:

	Activity Site	**Notice**
1	A licensed aerodrome or a site where a temporary aerodrome licence is required	60 days
2	An aerodrome or site when an aerodrome licence is unnecessary	28 days

2 If the aircraft commander becomes aware that he/she has contravened the regulation covered by **(1) c** above (i.e. national defence or affecting public interest), he/she is to leave the affected area by the shortest means and must not descend while over that area.

Balloons, kites, airships, gliders and parascending parachutes (ARTICLE 75)

1 Without the written permission of the CAA and subject to conditions granted by the CAA, and within the UK:

a A captive balloon or kite shall not be flown at a height of more than 60 m above the ground level or within 60 m of any vessel, vehicle or structure
b A captive balloon shall not be flown within 5 km of an aerodrome
c A balloon exceeding 2 m in any linear dimension at any stage of its flight, including any basket or other equipment attached to the balloon, shall not be flown in controlled airspace notified for the purposes of Article 75
d A kite shall not be flown within 5 km of an aerodrome
e An airship shall not be moored; and
f A glider or parascending parachute shall not be launched by winch and cable or by ground tow to a height of more than 60 m above ground level.

2 When a captive balloon is in flight it must be securely moored, and shall not be left unattended unless it is fitted with an automatic deflation device that operates if it breaks free of its moorings.

Aerodromes – public transport of passengers and instruction in flying (ARTICLE 76)

1 Aircraft that are described in paragraph (**2**) of this article must take off or land only at:

a An aerodrome licensed under the ANO for the take-off and landing of such aircraft; or
b A Government aerodrome, or a CAA owned or managed aerodrome, notified as available for the take-off and landing of such aircraft, or in respect of which the person in charge of the aerodrome has given their permission for the particular aircraft to take-off or land.

2 Aircraft for which paragraph (**1**) applies to, are:

a Aeroplanes of which the MTWA exceeds 2,730 kg and which are flying for public transport purposes, flying instruction purposes for qualifying for a licence or rating, flying test purposes for the grant of a pilot's licence or rating
b Aeroplanes of which the MTWA does not exceed, 2730 kg engaged on:
   i Scheduled journeys for public transport of passengers

ii Flights for the public transport of passengers beginning and ending at the same aerodrome
iii Flights for flying instruction purposes for qualifying for a licence or rating, or flying test purposes for the grant of a pilot's licence or rating
iv Night flights for the public transport of passengers

c Helicopters and gyroplanes flying as specified in sub-paras **(2) (b) (i)** and **(2) (b) (iii)** above; and
d Gliders flying for the public transport of passengers or flying instruction purposes (this does not apply to flying club gliders carrying club members only).

3 Whenever a helicopter is flying at night within the UK, for the public transport of passengers, with intent to land or take off from a place other than that described in paragraph **(1)**, the person in charge of that area must ensure that there is sufficient lighting in operation for the helicopter pilot to identify the landing area, determine the landing direction and to be able to make a safe approach and landing. The helicopter pilot must not make a landing or take off if he/she is flying with passengers for public transport purposes, unless the lighting is in operation.

Aviation fuel at aerodromes (ARTICLE 87)

3 If a person knows or has reason to believe that the aviation fuel is not fit for use in an aircraft, he/she must prevent that fuel from being dispensed for use to an aircraft. If the CAA or an authorised person suspects that unfit aviation fuel is intended or likely to be delivered to an aircraft, the CAA or that authorised person may direct the installation manager not to dispense aviation fuel from that installation until the direction has been revoked by the CAA or by an authorised person.

4 For the purpose of Article 87:

a 'Aviation fuel' means fuel intended for use in aircraft
b 'Aviation fuel installation' means any apparatus or container, including a vehicle, designed, manufactured or adapted for the storage of aviation fuel or for the delivery of such fuel to an aircraft.

Mandatory reporting (ARTICLE 94)

2 Reportable Occurrence means:

a Any incident relating to such an aircraft or any defect in or malfunctioning of such an aircraft or any part or equipment of such an aircraft, being an incident, malfunctioning or defect endangering, or which if not corrected would endanger, the aircraft, its occupants, or any other persons; and

b Any defect in or malfunctioning of any facility on the ground used or intended to be used for purposes of or in connection with the operation of such an aircraft, being a defect or malfunctioning endangering, or which if not corrected would endanger, such an aircraft or its occupants:

Provided that any notifiable accidents are reported as such and not as a reportable occurrence.

5 The operator of an aircraft shall, if he/she has reason to believe that a report has been or will be made, preserve any data from a flight data recorder or a combined cockpit voice recorder/flight data recorder relevant to the reportable occurrence for 14 days from the date on which a report of that occurrence is made to the CAA, or for a longer period if directed to do so by the CAA.

Provided that the record may be erased if the aircraft is outside the UK and it is not reasonably practicable to preserve the record until the aircraft reaches the UK.

The air navigation (general) regulations

Regulation 17 – reportable occurrences

For the purposes of Article 94, a Reportable Occurrence is one:

a Involving damage to an aircraft
b Involving injury to a person
c Involving the impairment during a flight of the capacity of a member of the flight crew of an aircraft to undertake the functions to which his/her licence relates
d Involving the use in flight of any procedures taken for the purpose of overcoming an emergency
e Involving the failure of an aircraft system or of any equipment of an aircraft
f Arising from the control of an aircraft in flight by its flight crew
g Arising from the failure or inadequacy of facilities or services on the ground used or intended to be used for the purposes of or in connection with the operation of aircraft
h Arising from the loading or the carriage of passengers, cargo (including mail) or fuel:

and any other occurrence that in the opinion of a person referred to in ANO Article 94(1), considers it an occurrence that endangers, or which if not corrected would endanger, the safety of an aircraft, its occupants or any other person.
The report shall contain:

a The type, series and registration marks of the aircraft concerned
b The name of the aircraft operator
c The date of the reportable occurrence
d If the person making the report has instituted an investigation into the reportable occurrence, whether or not this has been completed
e A description of the reportable occurrence, including its effects and any other relevant information
f In the case of a reportable occurrence which occurs during flight:
1 The Co-ordinated Universal Time of the occurrence
2 The last point of departure and the next point of intended landing of the aircraft at that time; and
3 The geographical position of the aircraft at that time

g In the case of a defect in or malfunctioning of an aircraft or any part or equipment of an aircraft, the name of the manufacturer of the aircraft, part or equipment, as the case may be, and, where appropriate, the part number and modification standard of the part or equipment and its location in the aircraft
h The signature and name in block capitals of the person making the report, the name of his/her employer and the capacity in which he/she acts for that employer; and
i In the case of a report made by the commander of an aircraft or a person referred to in sub-paragraphs (**c**) or (**d**) of Article 94 (1), the address or telephone number at which communications should be made to him/her, if different from that of that person's place of employment.

ABBREVIATIONS

AAL, aal	above aerodrome level
ABN	Aerodrome Beacon
A/c, a/c	Aircraft
ADA	Advisory Airspace
ADF	Automatic Direction Finder
ADR	Advisory Route
ADT	Approved Departure Time
AFIS	Aerodrome Flight Information Service
AFS	Aeronautical Fixed Service
AFTN	Aeronautical Fixed Telecommunications Network
agl	above ground level
AIC	Aeronautical Information Circular
AIP	Aeronautical Information Publication
AIREP	Air Report
AIS	Aeronautical Information Service
AMSL, amsl	above mean sea level
ANO	Air Navigation Order
ASR	altimeter setting region
ATAS	Air Traffic Advisory Service
ATC	Air Traffic Control
ATCC	Air Traffic Control Centre
ATCU	Air Traffic Control Unit
ATFM	Air Traffic Flow Management
ATIS	Automatic Terminal Information Service
ATS	Air Traffic Service
ATSU	Air Traffic Service Unit
Authority	Civil Aviation Authority
AUW	all-up weight
AWY	airway
C	degrees Celsius
CAA	Civil Aviation Authority
CANP	Civil Aircraft Notification Procedure
CAS	controlled airspace
CEU	Central Executive Unit (Flow Management)
CFO	Central Forecast Office
C of A	Certificate of Airworthiness
CTMO	Central Traffic Management Organisation
CTR	control zone
DA	decision altitude

DF	direction finding
DH	decision height
DME	distance measuring equipment
DR	dead reckoning
EAT	expected approach time
ECAC	European Civil Aviation Conference
ED	emergency distance
EET	estimated elapsed time
ELT	emergency location transmitters
EPIRB	emergency position indicating radio beacons
ETA	estimated time of arrival
F	degrees Fahrenheit
FAL	Facilitation of International Air Transport
FAX	facsimile transmission
FIR	Flight Information Region
FIS	Flight Information Service
FL	Flight Level
FLG	flashing
FMU	flight management unit
GASIL	General Aviation Safety Information Leaflet
GCA	ground controlled approach
H24	24 hour day/night service
HDG, Hdg	heading
HF	high-frequency (3,000–30,000 kHz)
Hr, hr	hour(s)
HZ, Hz	Hertz (cycle per second)
IAS	indicated air speed
IBN	identification beacon
ICAO	International Civil Aviation Organisation
ID	identification
IFR	instrument flight rules
ILS	instrument landing system
IMC	instrument meteorological conditions
ISA	International Standard Atmosphere
IRVR	instrument runway visual range
JAA	Joint Aviation Authorities
JAAR	JAA Requirements
kg	kilogram(mes)
kHz	kiloHertz
km	kilometre(s)
kt	knot(s)
LARS	Lower Airspace Radar Advisory Service
LATCC	London Air Traffic Control Centre
LDA	landing distance available
LF	low-frequency (30–300 kHz)
m	metre(s)
MATZ	Military Air Traffic Zone
mb	millibars
MEHT	minimum eye height over threshold
METAR	aviation routine weather report (in code)

MF	medium-frequency (300–3,000 kHz)
mHz	megaHertz
min	minute(s)
MNP	minimum navigation performance
MoD	Ministry of Defence
MOTNE	Meteorological Operational Tele-communications Network Europe
MTWA	maximum take-off weight authorised
MWO	Meteorological Watch Office
NAPs	noise abatement procedures
NATS	National Air Traffic Services
NDB	non-directional beacon
nm	nautical mile(s)
NOSIG	no significant change
OCA	Shanwick Oceanic Control Area
OCC	occluding (light)
OCL	obstacle clearance height
Order	Air Navigation Order
PANS	procedures for air navigation services
PAR	precision approach radar
PAX	passengers
RB	relative bearing
RCC	Rescue Co-ordination Centre
Regs	Regulations
RFP	repetitive flight plans
RIS	Radar Information Service
RLCE	request level change en route
RNAV	area navigation
RNLI	Royal National Lifeboat Institution
ROAR	Rules of the Air Regulations
RP	reporting point
R/T	radio-telephony
RTG	radio telegraphy (see W/T)
RTOW	regulated take-off weight
RVR	runway visual range
R/W, RWY	runway
SAR	search and rescue
SARP	ICAO Standards and Recommended Practices
SARSAT	Search and Rescue Satellite Aided Tracking System
Sch	Schedule
SELCAL	selective calling system
SFC	surface
SID	standard instrument departure
SIGMET	message of occurrence or expected occurrence of certain hazardous phenomena
SNOCLO	closed by snow
SNOWTAM	special message notifying of snow-affected movement areas
SPECI	aviation selected special weather report (in code)
SPL	supplementary flight plan message
SRA	Special Rules Area

SRZ	Special Rules Zone
SSR	secondary surveillance radar
STAR	standard instrument arrival
SVFR	special visual flight rules
TA/Trans Alt	transition altitude
TAF	aerodrome forecast
TAS	true air speed
TB	true bearing
TCA/TMA	Terminal Control Area
TODA	take-off distance available
TORA	take-off run available
TRLVL/ Trans Lev	transition level
TVOR	terminal VOR
TWR	aerodrome control or tower
UHF	ultra-high frequency
UIR	Upper Flight Information Region
UK	United Kingdom
u/s	unserviceable
UTC	co-ordinated universal time
VASIS	visual approach slope indicators
VFR	visual flight rules
VHF	very high frequency
VLF	very low frequency
VLR	very long range
VMC	visual meteorological conditions
VOR	very high frequency omni range
VSTOL	very short take-off and landing
VTOL	vertical take-off and landing
WEF	with effect from
WIP	work in progress
WPT	Waypoint
W/T	wireless telegraphy (see R/T)

DEFINITIONS

Aerial work means any purpose (other than public transport) for which an aircraft is flown if valuable consideration is given or promised in respect of the flight or the purpose of the flight provided that, if the only such valuable consideration consists of remuneration for the services of the pilot, the flight shall be deemed to be private flight for the purposes of Part III of the ANO.

Aerial work aircraft means an aircraft (other than a public transport aircraft) flying, or intended by the operator to fly for the purpose of aerial work.

Aerial work undertaking means an undertaking whose business includes the performance of aerial work.

Aerobatic manoeuvres includes loops, spins, rolls, bunts, stall turns, inverted flying and any other similar manoeuvre.

Aerodrome means any area of land or water designed, equipped, set apart or commonly used for affording facilities for the landing and departure of aircraft and includes any area or space, whether on the ground, on the roof of a building or elsewhere, which is designed, equipped or set apart for affording facilities for the landing and departure of aircraft capable of descending or climbing vertically, but shall not include any area the use of which for affording facilities for the landing and departure of aircraft has been abandoned and has not been resumed.

Aerodrome flight information unit means a person appointed by the Authority or by any other person maintaining an aerodrome to give information by means of radio signals to aircraft flying or intending to fly within the aerodrome traffic zone of that aerodrome, and **aerodrome flight information service** shall be construed accordingly.

Aerodrome operating minima in relation to the operation of an aircraft at an aerodrome means the cloud ceiling and runway visual range for take-off, and the decision height or minimum descent height, runway visual range and visual reference for landing, which are the minimum for the operation of that aircraft at that aerodrome.

Aerodrome traffic zone means the following specified airspace, being airspace in the vicinity of an aerodrome which is notified for the purposes of rule 35 of the Rules of the Air Regulations 1990(a):

a In relation to such an aerodrome other than one which is an offshore installation:

i At which the length of the longest runway is notified as 1,850 m or less, the airspace extending from the surface to a height of 2,000 ft above the level of the aerodrome within the area bounded by a circle centred on the notified mid-point of the longest runway and having a radius of 2 nm:

Provided that where such an aerodrome traffic zone would extend less than 1½ nm beyond the end of any runway at the aerodrome and this proviso is notified as being applicable, sub-paragraph (**ii**) following, shall apply as though the length of the longest runway is notified as greater than 1,850 m;

ii At which the length of the longest runway is notified as greater than 1,850 m, the airspace extending from the surface to a height of 2,000 ft above the level of the aerodrome within the area bounded by a circle centred on the notified mid-point of the longest runway and having a radius of 2½ nm

b In relation to such an aerodrome which is an offshore installation, the airspace extending from mean sea level to 2,000 ft above mean sea level and within 1½ nm of the offshore installation;

except any part of that airspace which is within the aerodrome traffic zone of another aerodrome which is notified for the purposes of the ANO as being the controlling aerodrome.

Aeronautical beacon means an aeronautical ground light which is visible either continuously or intermittently to designate a particular point on the surface of the earth.

Aeronautical ground light means any light specifically provided as an aid to air navigation, other than a light displayed on an aircraft.

Aeronautical radio station means a radio station on the surface, which transmits or receives signals for the purpose of assisting aircraft.

Air Traffic Control Clearance means authorisation by an air traffic control unit for an aircraft to proceed under conditions specified by that unit.

Air traffic control unit means a person appointed by the Authority or any other person maintaining an aerodrome or place to give instructions or advice or both instructions and advice by means of radio signals to aircraft in the interests of safety but does not include a person so appointed solely to give information to aircraft, and **air traffic control service** shall be construed accordingly.

Air transport undertaking means an undertaking whose business includes the carriage by air of passengers or cargo for valuable consideration.

Anti-collision light means:

a In relation to rotorcraft, a flashing red light.
b In relation to any other aircraft, a flashing red or flashing white light.

In either case showing in all directions for the purpose of enabling the aircraft to be more readily detected by the pilots of distant aircraft.

Approach to landing means that portion of the flight of the aircraft, when approaching to land, in which it is descending below a height of 1,000 ft above the relevant specified decision height or minimum descent height.

Apron means the part of an aerodrome provided for the stationing of aircraft for the embarkation and disembarkation of passengers, for loading and unloading of cargo and for parking.

Aviation fuel means fuel intended for use in aircraft.

Aviation fuel installation means any apparatus or container, including a vehicle, designed, manufactured or adapted for the storage of aviation fuel or for the delivery of such fuel to an aircraft.

Cabin attendant, in relation to an aircraft, means a person on a flight for the purpose of public transport carried for the purpose of performing in the interests of the safety of passengers, duties to be assigned by the operator or the commander of the aircraft, but who shall not act as a member of the flight crew.

Captive balloon means a balloon which when in flight is attached by a restraining device to the surface.

Cargo includes mail and animals and other freight.

Certificate of Airworthiness includes any validation thereof and any flight manual, performance schedule or other document, whatever its title, incorporated by reference in that certificate relating to the C of A.

Certificated for single-pilot operation means an aircraft which is not required to carry more than one pilot by virtue of any one or more of the following:

- **a** The C of A duly issued or rendered valid under the law of the country in which the aircraft is registered.
- **b** If no C of A is required to be in force, the C of A, if any, last in force in respect of that aircraft.
- **c** If no C of A is or has been previously in force but the aircraft is identical in design with an aircraft in respect of which such a certificate is or has been in force, the C of A which is or has been in force in respect of such an identical aircraft.
- **d** In the case of an aircraft flying in accordance with the conditions of a permit to fly issued by the Authority, that permit to fly.

Cloud ceiling in relation to an aerodrome means the vertical distance from the elevation of the aerodrome to the lowest part of any cloud visible from the aerodrome which is sufficient to obscure more than one-half of the sky so visible.

Commander in relation to an aircraft means the member of the flight crew designated as commander of that aircraft by the operator thereof, or, failing such a person, the person who is for the time being the pilot-in-command of the aircraft.

Congested area in relation to a city, town or settlement, means any area which is substantially used for residential, industrial, commercial or recreational purposes.

Controlled airspace means airspace which has been notified as Class A, Class B, Class C, Class D or Class E airspace.

Control area means controlled airspace which has been further notified as a control area and which extends upwards from a notified altitude or flight level.

Control zone means controlled airspace which has been further notified as a control zone and which extends upwards from the surface.

Copilot in relation to an aircraft means a pilot who is performing his duties and as such is subject to the direction of another pilot carried in the aircraft.

Crew means a member of the flight crew (a person carried on the flight deck who is appointed by the operator of the aircraft to give or to supervise the training, experience, practice and periodical tests required in respect of the flight crew as stated in the ANO), or a cabin attendant.

Danger area means airspace which has been notified as such within which activities dangerous to the flight of aircraft may take place or exist at such times as may be notified.

Day means the time from half an hour before sunrise until half an hour after sunset (both times exclusive), sunset and sunrise being determined at surface level.

Decision height in relation to the operation of an aircraft at an aerodrome means the height in a precision approach at which a missed approach must be initiated if the required visual reference to continue that approach has not been established.

Flight Crew in relation to an aircraft means those members of the crew of the aircraft who respectively undertake to act as pilot, flight navigator, flight engineer and flight radio operator of the aircraft.

Flight Level means one of a series of levels of equal atmospheric pressure, separated by notified intervals and each expressed as the number of hundreds of feet which would be indicated at that level on a pressure altimeter calibrated in accordance with the International Standard Atmosphere and set to 1013.2 mb.

Flight plan means such information as may be notified in respect of an air traffic control unit being information provided or to be provided to that unit, relative to an intended flight or portion of a flight of an aircraft.

Flight recording system means a system comprising either a flight data recorder or a cockpit voice recorder or both.

Flight simulator means apparatus by means of which flight conditions in an aircraft are simulated on the ground.

Flight visibility means the visibility forward from the flight deck of an aircraft in flight.

Free balloon means a balloon which when in flight is not attached by any form of restraining device to the surface.

Government aerodrome means any aerodrome in the UK which is in the occupation of any Government Department or visiting force.

Ground visibility means the horizontal visibility at ground level.

Hang glider means a glider capable of being carried, foot launched, and landed solely by the energy and use of a pilot's legs.

IFR flight means a flight conducted in accordance with the Instrument Flight Rules.

Instrument Meteorological Conditions means weather precluding flight in compliance with the Visual Flight Rules.

Licence includes any certificate of competency or certificate of validity issued with the licence or required to be held in connection with the licence by the law of the country in which the licence is granted.

Lifejacket includes any device designed to support a person individually in or on the water.

Manoeuvring area means the part of an aerodrome provided for the take-off and landing of aircraft and for the movement of aircraft on the surface, excluding the apron and any part of the aerodrome provided for the maintenance of aircraft.

Maximum total weight authorised in relation to an aircraft means the maximum total weight of the aircraft and its contents at which the aircraft may take-off anywhere in the world, in the most favourable circumstances in accordance with the C of A in force in respect of the aircraft.

Microlight aeroplane means an aeroplane having an MTWA not exceeding 390 kg, a wing loading at the MTWA not exceeding 25 kg/m^2, a maximum fuel capacity not exceeding 50 litres and which has been designed to carry not more than two persons.

Military aircraft means the naval, military or air force aircraft of any country and:

- **a** Any aircraft being constructed for the naval, military or air force of any country under a contract entered into by the Secretary of State; and
- **b** Any aircraft in respect of which there is in force a certificate issued by the Secretary of State that the aircraft is to be treated for the purposes of the ANO as a military aircraft.

Minimum descent height in relation to the operation of an aircraft at an aerodrome means the height in a non-precision approach below which descent may not be made without the required visual reference.

Nautical mile means the International Nautical Mile, that is to say, a distance of 1,852 m.

Night means the time from half an hour after sunset until half an hour before sunrise (both times inclusive), sunset and sunrise being determined at surface level.

Non-precision approach means an instrument approach using non-visual aids for guidance in azimuth or elevation but which is not a precision approach.

Notified means set forth in a document published by the CAA and entitled *United Kingdom Notam* or *United Kingdom Air Pilot* and for the time being in force.

Parascending parachute means a parachute which is towed by cable in such a manner as to cause it to ascend.

Passenger means a person other than a member of the crew.

Period of duty means the period between the commencement and end of a shift during which an air traffic controller performs, or could be called upon to perform, any of the functions specified in respect of a rating included in his/her licence.

Pilot-in-command in relation to an aircraft means a person who for the time being is in charge of the piloting of the aircraft without being under the direction of any other pilot in the aircraft.

Precision approach means an instrument approach using Instrument Landing System, Microwave Landing System or Precision Approach Radar for guidance in both azimuth and elevation.

Pressurised aircraft means an aircraft provided with means of maintaining in any compartment a pressure greater than that of the surrounding atmosphere.

Private flight means a flight which is neither for the purpose of aerial work nor public transport.

Public transport has a meaning assigned to it by Article 107 of the ANO.

Public transport aircraft means an aircraft flying, or intended by the operator of the aircraft to fly for the purpose of public transport.

Replacement in relation to any part of an aircraft or its equipment includes the removal and replacement of that part whether or not by the same part, and whether or not any work is done on it; but does not include the removal and replacement of a part which is designed to be removable solely for the purpose of enabling another part to be inspected, repaired, removed or replaced or cargo to be loaded.

Runway visual range in relation to a runway means the distance in the direction of take-off or landing over which the runway lights or surface markings may be seen from the touchdown zone as calculated by either human observation or instruments in the vicinity of the touchdown zone or where this is not reasonably practicable in the vicinity of the mid-point of the runway; and the distance, if any, communicated to the commander of an aircraft by or on behalf of the person in charge of the aerodrome as being the runway visual range shall be taken to be the runway visual range for the time being.

Scheduled journey means one of a series of journeys which are undertaken between the same two places and which together amount to a systematic service.

Seaplane has the same meaning as for the purpose of Section 97 of the Civil Aviation Act 1982.

Questions and Answers

Chapter 1 and Chapter 2 - Questions

1 The quadrantal rule applies:

- **a** Above the transition altitude
- **b** Above the transition layer
- **c** Above the transition level in notified airspace
- **d** Above the transition level outside controlled airspace

2 Automatic Terminal Information Service (ATIS) is:

- **a** Available on request
- **b** Subject to air traffic work load
- **c** Transmitted on a VHF frequency
- **d** Available at all major UK terminals

3 The manoeuvring area is:

- **a** All surfaces including the runways, taxiways, apron and maintenance areas
- **b** The taxiways, apron and maintenance areas
- **c** That part provided for take-off and landing, and movement of aircraft on the surface excluding the apron and the maintenance areas
- **d** The runways and taxiways only

4 An aircraft flying under IFR outside controlled airspace on a track of 359°M, must fly at the following pressure altitude:

- **a** FL 290
- **b** FL 330
- **c** FL 235
- **d** FL 310

5 An aircraft flying under VFR, at 4,000 ft AMSL and at 145 kt, must have a flight visibility of at least:

- **a** 1,800 m
- **b** 3 km
- **c** 5 km
- **d** 8 km

6 For flight under the instrument flight rules (IFR) inside controlled airspace:

- **a** An aircraft must maintain a clearance of at least 1,500 ft above the highest obstacle within 5 nm
- **b** An aircraft must maintain a clearance of at least 1,000 ft above the highest obstacle within 5 nm
- **c** An aircraft must maintain a clearance of at least 1,500 ft above the highest obstacle within 10 nm
- **d** An aircraft must maintain a clearance of at least 1,000 ft above the highest obstacle within 10 nm

7 The total width of an airway in the UK is:

- **a** 10 nm each side of the centreline
- **b** 15 nm
- **c** 5 nm each side of the centreline
- **d** 20 nm

8 An aircraft is cruising on an Advisory route, heading 275°M and experiencing 7° port drift. Which of the following would be a suitable level for this flight?:

- **a** FL 90
- **b** FL 65
- **c** FL 75
- **d** FL 80

9 Which flight level separates the FIR from the UIR?:

- **a** FL 245
- **b** FL 90
- **c** FL 10
- **d** FL 290

Chapter 3 – Questions

1 The stub of a Military Aerodrome Traffic Zone (MATZ):

- **a** Is 5 nm long, 4 nm wide with vertical extent between 1,000 ft AAL and 3,000 ft AAL
- **b** Is 4 nm long, 5 nm wide with vertical extent between 1,000 ft AAL and 3,000 ft AAL
- **c** Is 5 nm long, 4 nm wide with vertical extent between the surface and 3,000 ft AMSL
- **d** Is 4 nm long, 5 nm wide with vertical extent between 1,000 ft AMSL and 3,000 ft AMSL

2 Use of the Lower airspace Radar Advisory Service (LARS) is restricted to aircraft flying:

- **a** Below FL 100 with SSR transponders equipped with Mode C
- **b** Between the surface and FL 95
- **c** From military aerodromes below FL 95
- **d** Between 3,000 ft AMSL and FL 95

3 On the UK Airspace Restrictions chart, an area designated as restricted is defined by:

a A purple outline
b A pecked red outline
c A blue outline
d A pecked black outline

4 On the UK Airspace Restrictions chart, a danger area activated by NOTAM is defined by:

a A purple outline
b A pecked red outline
c A blue outline
d A pecked black outline

5 MATZ airspace is based on:

a 5 km radius from the centre of the longest runway, surface to 3,000 ft AAL
b 5 nm radius from the centre of the longest runway, surface to 3,000 ft AAL
c 5 nm radius of the ARP, surface to 3,000 ft AAL
d 5 nm radius of the ARP, surface to 3,000 ft AMSL

6 MATZ penetration requires contact to be established by:

a 15 nm, 5 min from the zone boundary
b 5 nm, 15 min from the zone boundary
c 15 min, 5 nm from the ARP
d 15 nm, 5 min from the ARP

7 In class D airspace a speed restriction exists under certain conditions. Which of the following statements are correct?:

a 140 kt speed limitation below FL100
b 140 kt speed limitation below 3,000 ft AMSL
c 250 kt speed limitation below FL100
d 250 kt speed limitation below 3,000 ft AMSL

8 When flying through a MATZ, the correct altimeter setting would be:

a QFE
b QNH
c QNE
d Regional QNH

Chapter 4 – Questions

1 A UK registered airship, while moored at night but not to a mooring mast, must display:

a A white light showing ahead, and a white light showing astern
b A white light astern only

- **c** A white light showing ahead, a white light astern and two red lights suspended at least 4 m below the control car
- **d** A steady white light astern, a steady red light on the port side and a steady green light on the starboard side

2 A captive balloon is required to show the appropriate lights providing it is:

- **a** More than 60 m AMSL
- **b** More than 150 m AGL
- **c** More than 60 ft AGL
- **d** More than 60 m AGL

3 A vehicle towing another aircraft converges from your left-hand side, while you are taxying. You should:

- **a** Maintain course and speed
- **b** Turn to the left
- **c** Slow down, or stop
- **d** Turn to the right

4 A glider overtaking another glider in flight:

- **a** Must overtake keeping the other glider on its right, and in sight at all times
- **b** Must overtake keeping the other glider to its left, and in sight at all times
- **c** May overtake on either side of the other glider, but keep it in sight at all times
- **d** Must do so by descending below the other glider

5 An aircraft registered in the UK before the 1st April 1988, with a maximum authorised weight of 5,600 kg, must display:

- **a** The standard prescribed navigation lights, with an anti-collision light
- **b** The standard prescribed navigation lights all flashing in alternation with an anti-collision light
- **c** The standard prescribed navigation lights, with a steady red anti-collision beacon
- **d** The standard prescribed navigation lights, with an anti-collision light optional

6 A glider while flying at night shall either show the navigation lights appropriate to a powered aircraft, or:

- **a** A steady red light showing in all directions
- **b** A flashing white light showing in all directions
- **c** A flashing red light showing in all directions
- **d** A steady white light showing in all directions

7 At an aerodrome without an ATC unit, you are on final approach to the runway behind another aircraft. Do you:

- **a** Wait until the runway is clear
- **b** Land a safe distance behind the other aircraft
- **c** Go around, straight along the runway
- **d** Land, keeping well clear to his right

8 A flying machine on a UK aerodrome shall display an anti-collision light:

- **a** By day or by night, when stationary on the apron with the engines running
- **b** By day or by night, immediately before departure
- **c** By day or by night, unless the standard navigation lights are displayed
- **d** By day or by night, unless stationary on the apron, or on an area designated for maintenance

9 In accordance with the General Flight Rule concerning avoidance of aerial collisions, an aircraft which is obliged to give way to another aircraft:

- **a** Must always do so by altering course to the right
- **b** Should manoeuvre in such a way to allow the other aircraft to maintain its course and speed
- **c** Should avoid passing over or under the other aircraft, unless passing well clear
- **d** Should maintain course and speed but slow down, allowing the other aircraft to pass well clear

10 One of the General Flight Rules states that an aircraft while landing, or on final approach to land, shall have the right-of-way over:

- **a** All other aircraft in flight, but not other aircraft on the manoeuvring area
- **b** All other aircraft on the manoeuvring area, but not all other aircraft in flight
- **c** Other aircraft in flight, or on the ground or water
- **d** All other aircraft flying in the aerodrome traffic zone

Chapter 5 – Questions

1 Aerodrome elevation is published in the *UK Air Pilot*, AGA section, and is:

- **a** The highest point on the landing area
- **b** The highest point on the aerodrome
- **c** The highest point on the manoeuvring area
- **d** The lowest runway threshold

2 The abbreviation MEDA means:

- **a** Maximum emergency distance available
- **b** Minimum emergency distance available
- **c** Military emergency diversion aerodrome
- **d** Minimum extra distance available

3 Emergency distance available is defined as:

- **a** The length of the runway
- **b** The length of the stopway plus the clearway
- **c** The length of the runway plus the stopway
- **d** The length of the runway plus the clearway

4 Unlicensed aerodromes are:

- **a** Not listed in any official CAA publication
- **b** Listed in the Air Navigation General Regulations
- **c** Listed in the *UK Air Pilot* AGA section, with all relative information verified by the Authority
- **d** Listed in the *UK Air Pilot* AGA section, but with no information verified

5 Information concerning the clearway for a particular runway, is for use by:

- **a** Departing aircraft only
- **b** Landing aircraft only
- **c** Aircraft making precautionary landings
- **d** Both departing and landing aircraft

6 The marshalling signal in which the right arm is held down, and the left arm is moved upwards and backwards, means:

- **a** Start the right engine
- **b** Turn to port, or open up the starboard engine
- **c** Stop
- **d** Turn to starboard

7 Runways exceeding 1,800 m in length and with no approach lighting system have fixed distance markers. These are:

- **a** 100 m from the threshold
- **b** 200 m from the threshold
- **c** 300 m from the threshold
- **d** 400 m from the threshold

8 Holding points are designated by what type of marks on the ground?:

- **a** Double solid and double pecked yellow lines
- **b** Single solid and single pecked yellow lines
- **c** Double solid and double pecked white lines
- **d** Single solid and single pecked white lines

9 TODA is:

- **a** TODA plus Clearway
- **b** TORA plus Clearway
- **c** EDA plus Clearway
- **d** EDA plus Stopway

10 At inactive military aerodromes, a white cross underlined by a white bar signifies:

- **a** The runway has been inspected within the last six months and is considered suitable for emergency use only
- **b** The runway is used for the storage of ammunition only and should not be used in any circumstances
- **c** It is considered by the authorities that the runway is fit enough for practice circuits to take place with caution
- **d** The runway is used for glider towing and other aircraft should remain clear of the area at all times

Chapter 6 – Questions

1 Aeronautical Information Circulars (AICs) are issued monthly and cover a wide range of subjects. After what period of time does a circular cease to be valid?:

- **a** 1 year
- **b** 2 years
- **c** 3 years
- **d** 5 years

2 An aeronautical information circular coloured yellow deals with:

- **a** Amendments to UK airspace restrictions
- **b** Matters with special emphasis on safety
- **c** Operational matters
- **d** Administrative matters and publications

3 Aeronautical Information Circulars are issued:

- **a** Only to flight training organisations
- **b** To any public transport operator
- **c** To Aeronautical Information Service (AIS) units only
- **d** To any subscriber

4 The AGA section of the UK Aeronautical Information Publication (CAP 32) deals with:

- **a** UK restricted airspace
- **b** ATC radio frequencies
- **c** UK aerodromes
- **d** General information on aeroplanes

5 A NOTAM is distributed by:

- **a** The AFTN and details temporary navigation warnings
- **b** Telex or Facsimile transmission
- **c** The AFTN and details changes of an operational significance that are introduced at short notice
- **d** Post, and details temporary navigation warnings

Chapter 7 – Questions

1 With reference to secondary surveillance radar, which of the following statements is correct?:

- **a** All aircraft flying in UK controlled airspace under IFR must carry a transponder with both Mode A and Mode C with altitude reporting
- **b** All aircraft flying in UK regulated airspace must carry a transponder with 4096 codes and Mode A. Mode C must be available for all flights above FL 100
- **c** Public Transport aircraft flying in UK regulated and advisory airspace must carry a transponder with 4096 codes and both Mode A and Mode C

- **d** All aircraft flying in UK regulated and advisory airspace must carry a transponder with 4096 codes and with Mode B available for use in emergency

2 If a pilot has been offered, and accepted, a Radar Advisory Service:

- **a** All instructions and vectors concerning conflicting traffic must be complied with
- **b** If necessary, altitude and/or heading may be changed without advising the controller
- **c** The pilot must have an instrument rating and the flight must be conducted in accordance with the Instrument Flight Rules
- **d** The controller will assume that the pilot will comply with instructions, unless the pilot advises otherwise

3 Under a Radar Information Service, separation from other traffic is the responsibility of:

- **a** The controller
- **b** The pilot
- **c** The pilot and the controller
- **d** The Senior Air Traffic Control Officer (SATCO)

4 If no other transponder code has been assigned, an aircraft engaged in the dropping of parachutists should select:

- **a** A4321 for the entire duration of the flight
- **b** A4321 five minutes before the drop commences until such a time that all the parachutists are on the ground
- **c** C7400 for the entire duration of the flight
- **d** C0033 five minutes before the drop commences until all the parachutists are estimated to be on the ground

5 Radar advisory service provides aircraft working the same unit with:

- **a** 3 nm horizontally between identified aircraft
- **b** 5 nm horizontally between identified aircraft
- **c** 1,000 ft vertically between identified aircraft
- **d** 1,000 ft vertically and 5 nm horizontally between identified aircraft

Chapter 8 – Questions

1 When the lower limit of a section of an airway is defined as a flight level, an absolute minimum safety altitude applies. That altitude is at least:

- **a** 1,500 ft above any fixed obstacle within 10 nm of the centreline
- **b** 1,000 ft above any fixed obstacle within 5 nm of the centreline
- **c** 1,500 ft above any fixed obstacle within 15 nm of the centreline
- **d** 1,000 ft above any fixed obstacle within 15 nm of the centreline

2 An aircraft may cross an airway without a clearance from ATC, only at:

- **a** Right angles, across the base of an en route sector
- **b** A TAS in excess of 150 kt
- **c** A reporting point
- **d** A reporting point and then at right angles

3 When authorised by NATS, radar separation standards below FL 245 may be reduced to 3 nm for identified aircraft flying within a certain distance of the radar head. The distance is:

- **a** 10 nm
- **b** 40 nm
- **c** 15 nm
- **d** 20 nm

4 A flight plan must be filed:

- **a** When a flight is made over sparsely populated or mountainous areas
- **b** If the planned flight crosses an FIR boundary
- **c** When a flight is made beyond 10 nm of the UK coast, and radio equipment is not carried
- **d** If a flight is to be made in Controlled Airspace in VMC, under the Instrument Flight Rules

5 When clearance to climb has been issued by ATC, the pilot may make vertical position reports in terms of flight levels:

- **a** Immediately after departure with an IFR clearance
- **b** As soon as the transition altitude has been passed
- **c** When the aircraft is no more than 2,000 ft below the transition altitude
- **d** As soon as the aircraft is within 2,000 ft of the transition level

6 If the base of an airway is defined as Flight Level, rather than an altitude:

- **a** The lowest usable level will provide at least 1,000 ft terrain clearance
- **b** That flight level may be used for cruising in accordance with the semi-circular rule
- **c** The lowest usable level will provide at least 1,500 ft terrain clearance
- **d** The lowest available flight level for cruising cannot be below FL 50

7 An obstacle defined as an 'Air Navigation Obstruction':

- **a** Is 300 ft AGL, or higher
- **b** Is 500 ft AGL, or higher
- **c** Is higher than 500 ft AMSL
- **d** Is higher than 500 ft AGL

8 ATC controllers are required to ensure that the levels assigned to an aircraft cruising on an airway provide a minimum terrain clearance. When the aircraft is more than 30 nm from the radar head, the minimum clearance is:

- **a** 1,000 ft above any fixed obstacle within 15 nm of the centreline
- **b** 1,000 ft above any fixed obstacle within 10 nm of the centreline
- **c** 1,500 ft above any fixed obstacle within 10 nm of the centreline
- **d** 1,500 ft above any fixed obstacle within 15 nm of the centreline

9 An inflight request to cross an airway should include the following:

- **a** Ident, aircraft type, heading, position, time of crossing airway
- **b** Ident, aircraft type, ETA at airway, point of crossing
- **c** Ident, aircraft type, position, heading, level, flight conditions and also requested position, level, and time of crossing
- **d** Ident, position, required track, level and also the requested point of crossing and the requested level

10 Longitudinal separation is based on time and distance. Which of the following statements is correct?:

- **a** Normal 15 min/10 nm, rapid fixing area 10 min
- **b** Rapid fixing area 10 min, with leading aircraft 40 kt faster then 2 min
- **c** Normal 10 min/15 nm, with leading aircraft 20 kt faster than 5 min/10 nm
- **d** Rapid fixing area 5 min, with leading aircraft 40 kt faster then 3 min

Chapter 9 – Questions

1 Any navigational aid that radiates the ident TST:

- **a** May be used as normal during the test transmission
- **b** Is permanently unserviceable
- **c** May be used with caution, but not for any approach procedure
- **d** Is transmitting for test purposes only and should not be used

2 A VOR transmitter is automatically shut down, and the standby transmitter activated, if omni-bearing information becomes less accurate than to:

- **a** 1 degree
- **b** 3 degrees
- **c** 5 degrees
- **d** 2 degrees

3 ILS localisers may be used to obtain bearings for position fixing en route. However, they should not be used:

- **a** At a range exceeding 25 nm, and are checked to a range of 10 nm
- **b** At a range exceeding 10 nm, to which they are fully checked and monitored
- **c** At a range exceeding 17 nm, and are checked to a range of 10 nm
- **d** At a range exceeding 17 nm, and are checked to a range of 15 nm

4 With reference to non-directional beacons:

- **a** They should not be used at night owing to unacceptable bearing errors
- **b** A designated operational coverage is published, outside which bearing error may be unacceptable
- **c** A minimum daytime protection is published, outside which unwanted signals may give bearing errors as great as 10 degrees
- **d** A minimum daytime protection ratio is published limiting bearing errors to 5 degrees

5 In the UK, ILS localiser transmitters provide coverage from the localiser aerial to distances of:

- **a** 25 nm within 10 degrees of the front course line, and 17 nm between 10 degrees and 35 degrees from the front course line
- **b** 10 nm within 35 degrees of the front course line, and 17 nm between 10 degrees and 35 degrees from the front course line
- **c** 25 nm within 10 degrees of the front course line, and 10 nm between 10 degrees and 35 degrees from the front course line
- **d** 20 nm within 10 degrees of the front course line, and 10 nm between 10 degrees and 35 degrees from the front course line

6 VOR bearing information is protected by a published value referred to as:

- **a** Promulgated range
- **b** Minimum protection ratio
- **c** Designated operational coverage
- **d** Daytime protection range

7 The SELCAL facility makes use of:

- **a** HF and VHF frequencies only
- **b** Any available VHF frequency
- **c** VHF and MF emergency frequencies
- **d** VHF and UHF emergency frequencies

8 With reference to VDF bearings, which of the following statements is correct?:

- **a** Class B bearings are accurate to within 5 degrees
- **b** Class D bearings are accurate to within 10 degrees
- **c** Class A bearings are accurate to within 1 degree
- **d** Class C bearings are accurate to within 1 degree

Chapter 10 – Questions

1 Which of the following is not included in VOLMET broadcast?:

- **a** QNH and RVR
- **b** RVR and dewpoint temperature
- **c** QFE and upper winds
- **d** Cloud base and visibility

2 If an aircraft has to divert to an aerodrome for which no meteorological forecast is provided, a request for the relevant meteorological information should be made to:

- **a** The air traffic control service which the aircraft is working
- **b** London VOLMET
- **c** The Meteorological Office at Bracknell
- **d** The air traffic control unit at the point of departure

3 The word SNOCLO is included in:

- **a** A Terminal Area forecast
- **b** An area forecast
- **c** A METAR
- **d** A VOLMET report

4 If meteorological information, including forecasts, is required for a flight of 500 nm, the request should be made:

- **a** 5 hr before departure
- **b** 3 hr before departure
- **c** 2 hr before departure
- **d** 1 hr before departure

5 If CAVOK is included in a meteorological broadcast:

- **a** There is no cloud at, or below 5,000 ft
- **b** There is no cloud below 5,000 ft or below the highest MSA, whichever is the greater
- **c** There is no cloud at or below 5,000 ft, or at or below the highest MSA, whichever is the greater
- **d** There is no cloud below 5,000 ft or below the highest MSA, whichever is the least

6 Which of the following meteorological conditions would not require a SIGMET to be issued, when cruising at subsonic levels?:

- **a** Cumulonimbus clouds
- **b** Severe icing
- **c** Tropical revolving storms
- **d** Marked mountain-wave activity

7 An IRVR system is used to measure runway visual range when:

- **a** The reported visibility is less than 1,000 m
- **b** The reported visibility is 1,000 m or less
- **c** The reported visibility is less than 1,500 m
- **d** The reported visibility is 1,500 m or less

8 At certain aerodromes, warning of a marked temperature inversion is given when:

- **a** A temperature difference of 10 degrees C or more exists between the surface and any point up to 1,000 ft AGL
- **b** The temperature at 1,000 ft AGL is 10 degrees C or more above the reported surface temperature
- **c** A temperature difference of more than 10 degrees C exists between the surface and any point up to 500 ft AGL
- **d** The temperature at 500 ft AGL is more than 10 degrees C above the reported surface temperature

9 In an IRVR system there are three transmissometers. If one of these units fails, then:

- **a** The system is rendered inoperative
- **b** The two remaining values are supressed, and the runway may not be used

- c The system continues to operate giving information from the two functioning units
- d The information will be supplied by a human observer using the standard RVR system

10 A gale warning is issued by a met observer:

- a When the mean surface wind exceeds 35 kt, or when gusts exceed 45 kt
- b When gusts exceed 45 kt
- c When the mean surface wind exceeds 33 kt or more, or when gusts exceed 42 kt
- d When the mean surface wind is 30 kt or more, or when gusts reach 45 kt or more

Chapter 11 – Questions

1 Which of the following emergency frequencies does not have a speech facility?:

- a 121.5 mHz
- b 500 kHz
- c 243 mHz
- d 2182 kHz

2 An aircraft using the frequency 2182 kHz to broadcast a distress message should transmit at:

- a The hour and hour + 30
- b The hour, hour + 15, hour + 30 and hour + 45
- c Hour + 10 and hour + 40
- d Hour + 15 and hour + 45

3 If the survivors from a crashed aircraft have left the visual signal X next to the wreckage, it signifies that:

- a No assistance is required and all survivors are proceeding to the nearest settlement
- b Assistance is required
- c It is not possible to land a rescue aircraft at the crash site
- d Medical assistance is required

4 If the survivors from a crashed aircraft have left the visual signal V next to the wreckage, it signifies that:

- a All passengers and crew have survived, and are proceeding in the direction shown
- b Severe injuries have been incurred and medical assistance is required
- c Assistance is required
- d No assistance is required

5 A listening watch is maintained at certain times on the international distress frequency 500 kHz. These times are:

- **a** H + 15 and H + 45 for 5 min
- **b** H and H + 30 for 5 min
- **c** H + 15 and H + 45 for 3 min
- **d** H and H + 30 for 3 min

Chapter 12 – Questions

1 If an aircraft inbound to the UK FIR is forced to land at an aerodrome other than a Customs aerodrome, the commander must:

- **a** Inform ATC and make arrangements for the flight to continue to a Customs aerodrome as soon as is possible
- **b** Inform the local Customs and Excise office within 30 min of landing
- **c** Await further instructions from the aircraft operator before departing for the next destination
- **d** Inform the local Customs Officer or the police immediately on landing

2 The commander of a scheduled flight, or his delegate, arriving at a UK Customs aerodrome from abroad, must provide:

- **a** One copy of the General Declaration
- **b** Two copies of the General Declaration
- **c** Three copies of the General Declaration
- **d** Four copies of the General Declaration

3 The commander of a scheduled flight, or his delegate, about to depart the UK for a foreign destination, must provide:

- **a** One copy of the General Declaration
- **b** Two copies of the General Declaration
- **c** Three copies of the General Declaration
- **d** No copies are required

4 The Air Navigation Order states that a foreign registered aircraft inbound to a UK aerodrome must not:

- **a** Commence an instrument approach if the minimum RVR is lower than that specified by the operator
- **b** Commence an instrument approach if the minimum RVR is lower than that specified by the authority
- **c** Commence an instrument approach if the cloudbase is lower than 200 ft above aerodrome level
- **d** Commence an instrument approach if the RVR and cloudbase are below those specified as minima by the authority

5 An aircraft flying to the UK from an aerodrome abroad may cross the UK coast-line:

- **a** Only on the specified cross-channel routeings
- **b** Only at the designated visual reporting points

c Only at the designated IFR fixing points or the visual reporting points
d At any point subject to airspace restrictions

Chapter 13 – Questions

1 The new owner of a UK registered aircraft must inform the CAA to this effect:

a Within 7 days
b Within 14 days
c Within 21 days
d Within 28 days

2 A weight and centre of gravity schedule is valid:

a For the duration of the relevant C of A
b For the time between two annual checks
c For a maximum of six months
d Indefinitely, unless equipment is added or removed

3 An aircraft has 126 passenger seats, but only 36 passengers on board. What is the minimum number of cabin attendants required to conform to the regulations?:

a One
b Two
d Three
d None

4 If a pilot attempts to renew his/her instrument rating three weeks before the expiry date, and fails the renewal flight test:

a The rating remains valid until the original expiry date
b The rating becomes invalid
c The rating is suspended and the following flight test must be conducted by a CAA staff examiner
d The rating is suspended pending further refresher training

5 A certain period of illness is allowed before a pilot is obliged to inform the CAA. This period is:

a 28 days
b 20 days
c Two months
d One month

6 If an aircraft operating for purposes of public transport requires that two pilots be carried:

a Both pilots must remain at the controls at all times for flight below FL 100
b Neither pilot is required to be at the controls during en route sectors, provided an approved autopilot is available
c Both pilots must be at the controls for take-off and landing
d The commander must wear safety straps at all times during the flight

7 In an unpressurised aircraft registered after 1st January 1989, and flying for the purpose of public transport:

- **a** The crew must use oxygen above FL 130
- **b** The crew must use oxygen above FL 100
- **c** It is recommended that all crew use oxygen above FL 130
- **d** It is recommended that all crew use oxygen above FL 150

8 In an unpressurised aircraft registered before 1st January 1989, and flying for purposes of public transport:

- **a** All passengers must be briefed on the use of the oxygen system before the flight reaches FL 130
- **b** All passengers and cabin attendants must use oxygen when the flight is above FL 100
- **c** All passengers must be briefed on the use of the oxygen system before the flight departs
- **d** No oxygen system is required if the flight does not reach and maintain levels above FL 120 for more than 30 min

9 When an aircraft is used for towing a glider:

- **a** The total length of the tow rope must not exceed 150 m
- **b** The total length of the aircraft, the tow rope and the glider must not exceed 200 m
- **c** The total length of the tow rope must not exceed 200 m
- **d** The total length of the aircraft, the tow rope and the glider must not exceed 150 m

10 How long must a personal flying log book be kept after the last entry date:

- **a** One year
- **b** Two years
- **c** Five years
- **d** Indefinitely

ANSWERS

Chapters 1 and 2

1 **a**, 2 **c**, 3 **c**, 4 **d**, 5 **c**, 6 **b**, 7 **c**, 8 **d**, 9 **a**

Chapter 3

1 **a**, 2 **b**, 3 **a**, 4 **b**, 5 **b**, 6 **a**, 7 **c**, 8 **b**

Chapter 4

1 **a**, 2 **d**, 3 **c**, 4 **c**, 5 **d**, 6 **a**, 7 **a**, 8 **a**, 9 **c**, 10 **c**

Chapter 5

1 **a**, 2 **c**, 3 **c**, 4 **d**, 5 **a**, 6 **b**, 7 **c**, 8 **a**, 9 **b**, 10 **a**

Chapter 6

1 **d**, 2 **c**, 3 **d**, 4 **c**, 5 **c**

Chapter 7

1 **a**, 2 **d**, 3 **b**, 4 **d**, 5 **c**

Chapter 8

1 **d**, 2 **a**, 3 **b**, 4 **d**, 5 **c**, 6 **c**, 7 **a**, 8 **d**, 9 **c**, 10 **b**

Chapter 9

1 **b**, 2 **a**, 3 **a**, 4 **d**, 5 **a**, 6 **c**, 7 **a**, 8 **a**

Chapter 10

1 **c**, 2 **a**, 3 **d**, 4 **c**, 5 **b**, 6 **a**, 7 **d**, 8 **a**, 9 **c**, 10 **c**

Chapter 11

1 **b**, 2 **a**, 3 **d**, 4 **c**, 5 **c**

Chapter 12

1 **d**, 2 **a**, 3 **d**, 4 **a**, 5 **d**

Chapter 13

1 **d**, 2 **d**, 3 **c**, 4 **b**, 5 **b**, 6 **c**, 7 **b**, 8 **a**, 9 **d**, 10 **b**

Index